Optimization-based Process Screening of Biorefinery Pathways at Early Design Stage

Optimierungsbasierte Prozessevaluierung von Bioraffineriepfaden zu einem frühen Entwicklungsstadium

Von der Fakultät für Maschinenwesen der Rheinisch-Westfälischen Technischen Hochschule Aachen zur Erlangung des akademischen Grades einer Doktorin der Ingenieurwissenschaften genehmigte Dissertation

vorgelegt von

Kirsten Svenja Mareike Skiborowski, geb. Ulonska

Berichter: Universitätsprofessor Alexander Mitsos, Ph. D.
Professor Antonis Kokossis, Ph. D.

Tag der mündlichen Prüfung: 21.09.2018

Aachener Verfahrenstechnik Series – Process Systems Engineering – Volume: 1 (2018)

Kirsten Skiborowski
Optimization-based Process Screening of Biorefinery Pathways at Early Design Stage
Optimierungsbasierte Prozessevaluierung von Bioraffineriepfaden zu einem frühen Entwicklungsstadium

ISBN: 978-3-95886-259-3

Bibliografische Information der Deutschen Bibliothek
Die Deutsche Bibliothek verzeichnet diese Publikation in der Deutschen Nationalbibliografie; detaillierte bibliografische Daten sind im Internet über http://dnb.ddb.de abrufbar.

Das Werk einschließlich seiner Teile ist urheberrechtlich geschützt. Jede Verwendung ist ohne die Zustimmung des Herausgebers außerhalb der engen Grenzen des Urhebergesetzes unzulässig und strafbar. Das gilt insbesondere für Vervielfältigungen, Übersetzungen, Mikroverfilmungen und die Einspeicherung und Verarbeitung in elektronischen Systemen.

Vertrieb:

    1. Auflage 2018
    © Verlagshaus Mainz GmbH Aachen
    Süsterfeldstr. 83, 52072 Aachen
    Tel. 0241/87 34 34 00
    www.Verlag-Mainz.de

Herstellung:

    Druckerei Mainz GmbH - Aachen
    Süsterfeldstraße 83
    52072 Aachen
    Tel. 0241/87 34 34 00
    www.DruckereiMainz.de

    Satz: nach Druckvorlage des Autors
    Umschlaggestaltung: Julia Kosloh

    printed in Germany
    D 82 (Diss. RWTH Aachen University, 2018)

Vorwort

Die vorliegende Arbeit entstand während meiner Zeit als wissenschaftliche Mitarbeiterin an der Aachener Verfahrenstechnik - Prozesstechnik und Systemverfahrenstechnik der RWTH Aachen. Mein besonderer Dank gilt meinem Doktorvater Professor Alexander Mitsos, Ph.D., sowie meinem Gruppenleiter Dr.-Ing. Jörn Viell für die Förderung und Unterstützung sowie die Offenheit gegenüber neuen Ansätzen und Ideen. Weiterhin bedanke ich mich bei Professor Dr.-Ing. Wolfgang Marquardt für die Betreuung und Unterstützung insbesondere während der ersten zwei Jahre meiner Promotion. Professor Antonis Kokossis, Ph.D., danke ich für die Übernahme des Koreferats und Professor Georg May, Ph.D., für die Übernahme des Prüfungsvorsitzes.

Weiterhin bedanke ich mich bei allen Mitarbeitern des Lehrstuhls für die kollegiale Zusammenarbeit sowie die freundschaftliche Atmosphäre am Lehrstuhl. Ich habe den fachlichen Austausch innerhalb der Synthese- und der SVT-TMFB-Gruppe als stets produktiv und sehr hilfreich wahrgenommen. Insbesondere bei Manuel Dahmen und Sebastian Recker möchte ich mich für ihre Hilfsbereitschaft und ihr kontinuierliches Feedback im Laufe der Jahre bedanken. Weiterhin danke ich allen Studierenden, die mich durch ihre Arbeit in Form von HiWi-Tätigkeiten, Lehrveranstaltungen und Abschlussarbeiten unterstützt haben. Die interdisziplinäre Zusammenarbeiten im Rahmen des TMFB haben mir geholfen meinen Horizont zu erweitern und meine Arbeit in einen größeren Kontext einzuordnen. Außerdem danke ich Daniel Maldonado, Andreas Harwardt und Caroline Marks für die gemeinsame Bürozeit, die ich sehr genossen habe. Manuel Dahmen, Preet Joseph Joy und Sebastian Recker danke ich für das Korrekturlesen der Arbeit. Ich bin außerdem dankbar für die vielen Freundschaften, die im Laufe der Lehrstuhlzeit entstanden sind.

Mein größter Dank gilt meiner Familie, insbesondere meinen Eltern, meinem Mann Mirko und unserem Sohn Joshua für Eure Unterstützung in jeder Lebenslage. Mit Eurer Liebe und Geduld habt Ihr einen großen Anteil am Gelingen dieser Arbeit.

Dortmund, im September 2018 *Kirsten Skiborowski*

Contents

Contents	**I**
Notation	**V**
Kurzfassung	**XII**
Summary	**XIII**
Publications and Copyrights	**XV**
1 Introduction	**1**
2 Screening approaches for biorefinery processes	**5**
2.1 Biorefineries	5
2.2 Process development	6
2.3 Process screening approaches for biorefineries	8
2.4 Reaction Network Flux Analysis	11
2.4.1 Network construction and input data	12
2.4.2 Model formulation	13
2.4.3 Shortcomings	17
2.5 Consideration of separation steps	17
2.6 Value chain analysis	19
2.6.1 Biomass supply chain	20
2.6.2 Market models	21
2.7 Uncertainties at early design stage	23
2.8 Requirements for a novel screening method	24
3 Process Network Flux Analysis	**27**
3.1 PNFA development	27
3.1.1 Flux balance	31

		3.1.2	Input data	32
	3.2	\multicolumn{2}{l}{Evaluation of processing pathways}	33
		3.2.1	Energy demand of pathways	34
		3.2.2	Economic efficiency of pathways	43
		3.2.3	Sustainability of pathways	48
		3.2.4	Allocation	53
	3.3	\multicolumn{2}{l}{Supply chain design}	55
	3.4	\multicolumn{2}{l}{Market and price modeling}	58
	3.5	\multicolumn{2}{l}{Optimization problem formulation}	60
		3.5.1	RNFA problem	60
		3.5.2	PNFA for processing pathways	61
		3.5.3	PNFA for value chains	62
	3.6	\multicolumn{2}{l}{Summary and conclusions}	64

4 PNFA benchmarking with literature, RNFA and conceptual design results — 65

	4.1	Cost and GWP comparison of PNFA and literature		66
		4.1.1	Comparison of minimum selling prices	66
		4.1.2	Comparison of investment costs	68
		4.1.3	Comparison of global warming potential	70
	4.2	\multicolumn{2}{l}{Accuracy of the PNFA results for an ethanol production compared to RNFA and conceptual design}	70
		4.2.1	Conceptual process design for ethanol production	71
		4.2.2	Conceptual design results	73
		4.2.3	Comparison of RNFA, PNFA and conceptual design results	75
	4.3	\multicolumn{2}{l}{Summary and conclusions}	76

5 Application of PNFA — 79

	5.1	Single-product biorefinery for fuel production		80
		5.1.1	Reaction network	80
		5.1.2	RNFA results	82
		5.1.3	Analysis of separations	83
		5.1.4	PNFA results	85
		5.1.5	Sensitivity analysis	94
		5.1.6	Heat integration	97
		5.1.7	Biomass supply chain design	99
	5.2	\multicolumn{2}{l}{Analysis of fuel mixtures}	107
	5.3	\multicolumn{2}{l}{Multi-product biorefinery for a co-production of fuel and chemicals}	114
	5.4	\multicolumn{2}{l}{Summary and conclusions}	116

6 Biorefinery improvement potential 119
  6.1 Key improvement factors 119
  6.2 Potential of biotechnology conversions 123
    6.2.1 Process screening 125
    6.2.2 Future improvement potential 129
    6.2.3 Summary 130

7 Conclusions and outlook 131

A Separation models 137

B Green chemistry metrics 141

C Case study parameters 144
  C.1 Overview parameters 144
  C.2 Reaction network 147
  C.3 Energy demand of separations 149
  C.4 Properties 153
  C.5 Active pathway fluxes 155
  C.6 Supply chain data and design 158
  C.7 Multi-product biorefinery 162
  C.8 Mixtures 163
  C.9 Fermentation data 163
  C.10 Conceptual design data 170

Bibliography 173

Notation

Latin symbols

A	matrix of stoichiometric coefficients	[-]
AE	atom economy	[-]
AU	atom utilization	[-]
b	product vector	[kmol/yr]
BTC	specific biomass transportation cost	[US $/yr]
c	concentration	[g/L]
Cap	capacity	[ton/yr]
CAGR	component annual growth rate	[%]
CE	carbon efficiency	[-]
CED	cumulative energy demand	[MJ/kg]
CED_{fuel}	CED for fuel production	[MJ/MJ]
cp	specific heat capacity	[kJ/kgK]
D	distance	[km]
dia	diameter	[m]
e	primary energy factors	[-]
E	energy demand	[kJ/yr]
$\mathbf{E_{spec}}$	specific energy demand	[kJ/mol]
ex_x^0	standard molar exergy of element X	[kJ/mol]
EC	energy efficiency combustion	[-]
EE	element efficiency	[-]
Eform	energy efficiency formation	[-]
EI	lumped environmental impact	[-]
Em	emissions	[mol CO_2equivalents/mol product]
EMY	effective mass yield	[-]
E_{ex}	exergy of a component	[kJ/mol]
f	flux vector	[kmol/yr]

v

Notation

F	factor	[-]
$\Delta_f G^0$	Gibbs free energy of a molecule	[kJ/mol]
gRME	generalized reaction mass efficiency	[-]
gwp	global warming potential factors	[-]
GWP	global warming potential	[g CO_2equivalents/MJ]
h	enthalpy	[kJ/kmol]
H	height	[m]
HC	non-accumulated heat cascade	[kJ/yr]
HC$_{sum}$	accumulated heat cascade	[kJ/yr]
HC$_{final}$	final heat cascade	[kJ/yr]
HI	matrix heat integration	[kJ/kmol]
HS	Hildebrandt solubility	[$MPa^{0.5}$]
IC	investment cost	[US $]
Inv1	pre-factor investment cost correlation	[-]
Inv2	exponent investment cost correlation	[-]
ir	interest rate	[%]
i	control variable	[-]
j	control variable	[-]
k	control variable	[-]
$\log K_{OW}$	octanol-water partition coefficient	[-]
M	molar mass	[kg/kmol]
MI	mass intensity	[-]
ms	market size vector	[tons/yr]
MSP	minimum selling price	[$/L]
MPF	Guthrie factor for operating conditions	[-]
MF	Guthrie factor for type of construction	[-]
$\dot{m}$	mass flow	[kg/yr]
N	number	[-]
NFU	number of functional units	[-]
n	run time	[yr]
$\dot{n}$	molar flow	[kmol/yr]
OE	optimum efficiency	[-]
p	pressure	[bar]
p_{oi}	vapor pressure	[atm]
pKs	acid dissociation constant	[-]
P	price	[$/kg]
P_{fuel}	fuel price	[$/L]
PMI	process mass intensity	[-]
R	ideal gas constant	[kJ/kmol K]
Ret	retention	[-]
RI	renewables intensity	[-]

RC	resource consumption	[-]
RME	reaction mass efficiency	[-]
RP	renewable percentage	[-]
SI	solvent intensity	[-]
SER	solvent energy ratio	[-]
Sol	solubility in water	[g/L]
Sol_1	pre-factor solubility	[g/L]
Sol_2	temperature-dependent solubility factor	[1/degree C]
SR	solvent recovery energy	[kJ/kg]
STY	space time yield	[kg/m^{3}s]
t	time	[h]
T	temperature	[K]
T_B	boiling point temperature	[K]
T_{melt}	melting point temperature	[degree C]
TAC	total annualized cost	[US \$/yr]
TBTC	total biomass transportation cost	[US \$/yr]
ToxP	toxicity potential	[-]
ToxS	toxicity score	[yr/ kg]
TP	total profit	[US \$/yr]
TR	total revenues	[US \$/yr]
V	volume	[m^3]
V_M	molar volume	[cm^3/mol]
VR	logistic factor "Verkehrsrundschau"	[-]
w	weight fraction	[-]
wc	weighting criteria EI	[-]
WI	waste intensity	[-]
WP	waste percentage	[-]
WER	waste energy ratio	[kg/kJ]
x	mole fraction	[-]
X	conversion	[-]
y	integer variable	[-]
Y	yield coefficient	[-]

Notation

Greek symbols

α	design target	[kJ/yr]
γ	polytropic exponent	[-]
ζ	split fraction boiler/ turbine	[-]
η	efficiency	[-]
κ	dissociation degree	[-]
λ	number of plant sites	[-]
ν	stoichiometric coefficient	[-]
ρ	density	[kg/m^3]

Subscripts and superscripts

BT	biomass transportation
chem	chemical
comb	combustion
comp	compressor
cool	cooling water
d	destinations
DS	downstream processing
elec	electricity
enz	enzymes
fus	fusion
GPP	gas phase processes
heat	heat demand
HEX	heat exchanger
HI	heat integration
IES	internal energy supply
i	substance
j	reaction
kr	key reactant
LV	vaporization
m	mixing
ML	mother liquor
mp	main product
mt	material
p	pressure

PV pressure vessel
o   origins
r   reaction
s   supply
SL melting
t   separation
T   temperature
TI temperature interval
var variable
w   waste residues

Abbrevations

ABE	acetone, butanol, ethanol
AVT	Aachener Verfahrenstechnik
BTHF	butyltetrahydrofuran
C	component
CAGR	component annual growth rate
CED	Cumulative Energy Demand
CEPCI	chemical engineering plant cost index
CH4	methane
COSMO-RS	Conductor-like Screening Model for Realistic Solvents
CO2	carbon dioxide
Cryst	cooling crystallization
CSTR	continuous stirred tank reactor
DCM	di-chloromethane
DNBE	di-n-butylether
DMSO	di-methylsulfoxid
DOE	di-octylether
ED	electrodialysis
EL	ethyllevulinate
EOE	ethyloctylether
EOS	equation of state
Evap	thermal separation
Extract	extraction
FA	furfuralacetone
GBL	gamma-butyrolactone
GVL	gamma-valerolactone

Notation

GWP	global warming potential
HD	heteroazeotropic distillation
HMF	hydroxymethylfurfural
HPA	3-hydroxypropionic acid
H2	hydrogen
LA	levulinic acid
LCA	life-cycle analysis
LCE	life-cycle energy
MILP	mixed-integer linear programming
MINLP	mixed-integer non-linear programming
MS	Marshall-Swift index
MTHF	methyltetrahydrofuran
NA	not available
NRTL	non random two liquid
NRW	North-Rhine Westphalia
OAT	one-at-a-time (analysis)
OL	1-octanol
PB	pollution, bioaccumulation
PFR	plug-flow reactor
PNFA	Process Network Flux Analysis
Pre	precipitation
PSD	pressure swing distillation
PTHF	propyltetrahydropyran
RX	reaction number X
RNFA	Reaction Network Flux Analysis
RS	reaction solvent
SHCF	separate hydrolysis and co-fermentation
SSCF	simultaneous saccharification and co-fermentation
THFA	4-(2-tetrahydrofuryl)-2-butanol
TON	turnover number catalyst
VLE	vapor liquid equilibrium
VLLE	vapor liquid liquid equilibrium
VR	Verkehrsrundschau Index
VRC	vapor recompression
WT	waste treatment energy

Notation

Kurzfassung

Für die Erhöhung der Nachhaltigkeit chemischer Prozesse spielt der Wandel von konventionellen zu erneuerbaren Ressourcen eine große Rolle. Das eröffnet eine Vielzahl an neuen Prozessvarianten. Eine detaillierte Auslegung aller Varianten ist aufwändig und teuer, da die erforderlichen Simulationen stark von Designparametern abhängen und es kommerzieller Simulationssoftware an Robustheit mangelt. Daher sind Screeningmethoden erforderlich zur Bewertung neuer Prozesse. Existierende Methoden sind auf die Analyse von Reaktions- oder literaturbekannter Prozesspfade beschränkt, sodass neue Pfade Simulationsstudien erfordern, was durch eine geringe Datenverfügbarkeit zu den Reaktionen und dem Fehlen fundierter Stoffdatenmodelle erschwert wird.

Um die Prozessentwicklung und -verbesserung zu beschleunigen, wird die Prozessnetzwerkflussanalyse als optimierungsbasierte Screeningmethode vorgestellt. Diese erfasst systematisch die Reaktionsdaten und identifiziert die Art, Machbarkeit und Effizienz von Trennungen, welche anhand thermodynamisch valider Trennmodelle evaluiert werden. Basierend auf Massen- und Energiebilanzen, werden die Pfade hinsichtlich ihrer Nachhaltigkeit und Wirtschaftlichkeit bewertet. Des Weiteren erlaubt die Methode eine Abschätzung des Wärmeintegrationspotentials, berücksichtigt die Biomassetransportkette und identifiziert vielversprechende Produktportfolios anhand eines pragmatischen Marktmodells. Somit ist die Methode für einzelne, mehrere parallele Produkte sowie Mischungen anwendbar.

Die Genauigkeit der Resultate wird anhand eines Vergleiches mit der Literatur, der Reaktionsnetzwerkflussanalyse und einer konzeptionellen Designstudie analysiert. Für eine komplexe Studie zur Produktion von Kraftstoffen aus Biomasse, wird die Anwendbarkeit demonstriert. Dabei ist die Produktion von Ethanol am vielversprechendsten, gefolgt von iso-Butanol. Eine profitable Produktion, insbesondere unter Berücksichtigung des Biomassetransports, ist nicht möglich. Diese wird nur durch eine Ko-Produkten von Chemikalien erzielt. Schlüsselfaktoren zur Entwicklung effizienter Bioraffinerien werden daraus abgeleitet. Dafür wird insbesondere das aktuelle sowie das theoretische Potential selektiver biotechnologischer Konversionen diskutiert.

Summary

In order to increase sustainability of chemical processes, a raw material change from conventional to renewable feedstocks is the key. This opens up numerous novel process concepts. A detailed conceptual design of all of these different pathways is expensive and time-consuming, since the mandatory simulations depend on pre-specified design decisions and commercial simulation software lack robustness. Hence, screening methodologies are required for an initial assessment of the processes. Existing screening methods are restricted to reaction or process design data known in literature, such that the integration of novel pathways requires simulation studies. This is impeded by limited data availability and the lack of profound property models.

Process Network Flux Analysis is introduced as an optimization-based screening methodology to accelerate process development and improvement for existing and novel processes. The method systematically integrates reaction data with the selection of separation processes, the feasibility and efficiency of which are evaluated using thermodynamically-sound separation models. Based on mass and energy balances, the pathways are analyzed according to their economic efficiency and sustainability. Furthermore, the method allows for an initial heat integration potential analysis, considers the influence of the biomass supply chain and identifies suitable product portfolios based on a pragmatic market model. Thus, the method is applicable for single or multiple products as well as for mixtures.

The accuracy of the results is analyzed by a comparison with literature data, Reaction Network Flux Analysis and a conceptual design study. For a complex case study of fuel production from biomass, the applicability of the method is demonstrated. The production of ethanol is most promising, followed by iso-butanol. A profitable production is not achieved, in particular, when the biomass supply chain is included. Profitability is only obtained by a co-production of chemicals. Finally, key improvement factors for future biorefineries are derived. For this purpose, the actual and theoretical potential of selective biotechnological conversion are discussed.

Publications and Copyrights

This thesis originates from the research performed by the author during her time at the Chair of Process Systems Engineering at Aachener Verfahrenstechnik (AVT). Most parts of this thesis have already been published. All publications are published under the author's birth name Kirsten Ulonska. The publications are integrated in the following chapters as described:

- Chapter 2: The literature overview presented in Section 2.3 is partially published in Ulonska et al. (2016a) and Ulonska et al. (2018). The latter article also includes the work in Section 2.6. Parts of the RNFA description (Section 2.4) and the analysis of uncertainties (Section 2.7) are published in Ulonska et al. (2016b).

- Chapter 3: In Ulonska et al. (2016a) a general PNFA concept (Section 3.1), parts of the evaluation criteria (Sections 3.2.2, 3.2.2.2, 3.2.3) as well as the optimization problem for the analysis of processing pathways (Section 3.5.3), have been presented. A first attempt to a systematic evaluation of fermentation setups is described in Ulonska et al. (2015), which is further developed in this thesis in Section 3.2.1.2. The PNFA coupled with supply chain design and market modelling (Sections 3.3, 3.4, 3.5.3) are published in Ulonska et al. (2018).

- Chapter 4: Section 4.1.1 and 4.1.3 are published in Ulonska et al. (2018).

- Chapter 5: The case study in Section 5.1 has been presented in Ulonska et al. (2016a). In addition to the results in the publication, the PNFA results herein include cost contributions from the utilities, which alter the results of internal energy supply. Furthermore, a detailed discussion of the cost structures as well as a comparison of a simultaneous versus sequential heat integration is added. The extension of the case study to include the supply chain (Section 5.1.7) and market model (Section 5.3) has been published in Ulonska et al. (2018).

- Chapter 6: Parts of the fermentation screening (Section 6.2) has been published in Ulonska et al. (2015) and is further refined in this thesis.

Publications and Copyrights

Additional contributions to publications are made, which are not part of this thesis. In Ulonska et al. (2016b), the RNFA methodology is applied to a large biofuel case study, but the results are not shown herein. In König et al. (2018) the PNFA methodology is utilized for a performance comparison of bio-fuels and fuels derived using renewable electricity, which are not part of this thesis as well.

While at the AVT, the author supervised the student theses of Klatt (2013), Bomheur (2013), Ramesh (2013), Gerdes (2014), Kochs (2015), Beckmann (2015), Rekkas Ventiris (2015), Gantner (2015), Vellguth (2016), Monigatti (2016), Wloemer (2016) and König (2016). These are developed as a cooperation between the author and the students. The work of all students is acknowledged. The following student theses are described partially in this work. The discussion on the problems regarding an automated network download from Reaxys are based on the experiences gained in the thesis of Gantner (2015). The supply chain and market submodel have been developed in the thesis of König (2016). The conceptual design study has been developed in the thesis of Wloemer (2016). An initial screening of fermentation processes has been conducted by Gerdes (2014), whereas an initial version of the precipitation model has been developed in the thesis of Kochs (2015).

Copyrights

Parts of the following publications, especially figures, are included in this thesis and are reprinted with permission:

- Reprinted from Ulonska, K., Ebert, B. E., Blank, L. M., Mitsos, A., and Viell, J. (2015). Systematic screening of fermentation products as future platform chemicals for biofuels. *Computer Aided Chemical Engineering*, 37:1331-1336. Copyright © (2018), with permission from Elsevier.
- Reprinted from Ulonska, K., Skiborowski, M., Mitsos, A., and Viell, J. (2016a). Early-stage evaluation of biorefinery processing pathways using process network flux analysis. *AIChE Journal*, 62(9):3096-3108. Copyright © (2018) with permission from Wiley.
- Reprinted with permission from Ulonska, K., Voll, A., and Marquardt, W. (2016b). Screening pathways for the production of next generation biofuels. *Energy & Fuels*, 30(1):445-456. Copyright © (2018) American Chemical Society.
- Reprinted with permission from Ulonska, K., Koenig, A., Klatt, M., Mitsos, A., and Viell, J. (2018). Optimization of multi-product biorefinery processes under consideration of biomass supply chain management and market developments. *Industrial & Engineering Chemistry Research*, 57(20):6980-6991. Copyright © (2018) American Chemical Society.

Chapter 1

Introduction

The implementation of competitive and sustainable chemical processes requires a raw material change from conventional to renewable resources in order to reduce greenhouse gas emissions as well as the exploitation of fossil feedstocks. As one of the most abundant renewable feedstocks, the processing of biomass is of particular interest and has resulted in the proposal of a high number of biomass processing concepts. In combination with a comprehensive product design, these concepts offer the potential for novel and superior pathways and products. However, in order to establish such novel pathways and products in the industrial value chain, a competitive production is mandatory.

The development of such processes is extremely complex, as not only a variety of biomass conversion technologies, but also a multitude of obtainable products have been proposed. Even if biomass processing is considered only, e.g., by specifying product properties or considering a predefined list of product candidates, a large and complex design problem for the identification of the most promising pathways remains. While in the petrochemical industry systematic approaches for conceptual process design are established, a direct transfer to the design of biorefineries is challenging. One reason is the limited availability of data, as often only proof-of-concept laboratory data exist without information about the reaction kinetics. For novel and highly functional molecules, physical property data is scarce. In addition, multiple pathways, which are composed of different reactions, can exist for the production of a specific product. Therefore, process simulations of the different opportunities are tedious and, thus, not practicable. Consequently, screening approaches are required, which rely on first principle data. The screening accelerates the detection of bottlenecks, identifies improvement potential and enables a focus on the most promising approaches at an early stage in process development.

1 Introduction

For this purpose, several approaches exist in literature (e.g., Bao et al., 2011; Santibanez-Aguilar et al., 2011; Rangarajan et al., 2014a; Cheali et al., 2014; Kelloway and Daoutidis, 2014; Zondervan et al., 2011; Celebi et al., 2017). All of these screening approaches rely on a superstructure formulation, which covers different process alternatives. Discrete variables indicate the presence or absence of an alternative. The resulting optimization problems target the minimization of the energy demand or global warming potential as well as the production costs (Kim et al., 2013; Wang et al., 2013; Rizwan et al., 2015; Quaglia et al., 2015; Garcia and You, 2015b; Gargalo et al., 2016a,b). All of these screening methods are based on known processes from literature or rely on elaborate and extensive simulation studies, which often lack robustness. These approaches are tedious, such that preparation times of more than a year are described (Kokossis et al., 2015).

For the quick assessment of existing and novel pathways based on limited laboratory data, Voll and Marquardt (2012a) proposed the Reaction Network Flux Analysis (RNFA) as an optimization-based screening method. RNFA determines the economic efficiency and sustainability of reaction pathways based on mole balances, assuming ideal separations. While a first insight into a large number of reaction pathways is enabled, the influence of reaction solvents on the choice, feasibility and effort of separations are neglected. Since the processes depend strongly on the separation effort, separations need to be considered simultaneously and addressed as early as possible.

In order to resolve these decisions, this thesis proposes an extended methodology, which is termed Process Network Flux Analysis (PNFA). PNFA bridges the gap between RNFA and conceptual design. Therefore, a biorefinery process framework for the assessment of novel processes is proposed, which is inspired by the work of Kraemer et al. (2009) and Skiborowski et al. (2013) for the development of optimal separation tasks. The biorefinery process framework gradually increases the model complexity from RNFA to PNFA and conceptual design by simultaneously reducing the number of process alternatives. Both, the model formulations of the RNFA and PNFA can be interpreted as superstructures as well, such that in principle any alternative screening method could have been used as starting point for this thesis. However, these alternative screening methods rely on input data from the conceptual design step and thus, do not support the biorefinery process framework for novel processes. Although these methods could have been adapted, the RNFA is preferred as starting point to maintain a similar input format and model for the first two steps of the biorefinery process framework. Herein, the RNFA is further developed to overcome the described limitations.

Therefore, PNFA is developed to systematically integrate and address separations, considering the various solvents along a pathway. Thus, PNFA is capable of evaluating the sustainability and economic efficiency for a multitude of interlinked novel reaction and processing steps. The process concepts are ranked based on the results of a multi-objective optimization, which provides a more detailed and clear analysis of process performance compared to the RNFA. The pathways can be ordered according to their economic efficiency or sustainability in terms of cumulative energy demand or global warming potential. Thereby, process bottlenecks are detected and, thus, pathway improvement potential is identified. In order to improve the quality of the results further, an analysis of the potential for heat integration and a biomass supply chain design coupled with a pragmatic market model are integrated in the PNFA. The latter takes into account price developments subject to supply and demand principles to evaluate the effort of a newly introduced production process on obtainable market prices. The method can evaluate single- and multi-product plants as well as product mixtures.

At first, a short elaboration on biorefinery concepts is given in Chapter 2 to set the contribution of this thesis into context. A summary of existing work on process screening methods, biomass supply chain design and market modeling is presented. PNFA, its fundamentals and mathematical model are then described in Chapter 3. Besides, material and energy balances, economic and sustainability criteria are addressed, emphasizing the extended analysis opportunities compared to the RNFA. In order to evaluate the accuracy of the PNFA results, a comparison with available processes in literature, RNFA and a conceptual design case study for ethanol is conducted in Chapter 4. In Chapter 5, the full potential of the method is illustrated for a complex case study covering all described aspects to obtain viable process concepts. Chapter 6 discusses the results for the development of key improvement factors for biorefineries and specifically addresses the potential of biotechnological conversions. Finally, Chapter 7 summarizes the innovative and selective aspects of the PNFA methodology and presents an outlook on further improvements.

Chapter 2

Screening approaches for biorefinery processes

The chapter starts with a general overview on biorefinery concepts before briefly introducing important work in the context of process design. Then, existing work in the field of process screening approaches for biorefineries is summarized. Most of these screening methods are restricted to existing designs or literature data. In order to analyze novel processes as well, Reaction Network Flux Analysis is introduced as an optimization-based screening method for the analysis of reaction pathways. The focus is on RNFA since the proposed methodology builds upon this. Therefore, the general RNFA idea, model formulation as well as current shortcomings are discussed in more detail. In order to expand the analysis to the assessment of full value chains, relevant literature in the fields of biomass supply chain design and market modelling is presented. The chapter ends with a summary on the requirement of a novel screening method.

2.1 Biorefineries

A biorefinery is defined as *"an overall concept of a processing plant where biomass feedstocks are converted and extracted into a spectrum of valuable product"* or more specifically as *"a facility that integrates biomass conversion processes and equipment to produce fuels, power, and chemicals from biomass"* (Kamm et al., 2006). In this thesis, a slightly adapted biorefinery definition is used such that fuel production from biomass is the main focus, whereas the benefit from a co-production of power or chemicals is discussed additionally.

The biomass feedstock might be starchy biomass with a low lignin content like corn, sugar cane, sugar beet or palm oil. The resulting biofuels are named first generation biofuels, with ethanol as most prominent example (Alonso et al., 2010). The annual ethanol production in 2016 of 79 million tons is twice as high compared to 2007 (Renewable Fuels Association, 2017). Although the global production capacity is still rising, these biofuels are criticized, since the feedstock could be used for feed or fodder production instead. Furthermore, jungle area is chopped down and used for palm plantation (Graham-Rowe, 2011). Hence, the conversion of lignocellulosic biomass as most abundant biomass resource is targeted, e.g., converting forestry and agricultural residues as well as energy crops. In addition, lignocellulosic biomass does not compete with feed or fodder production. The lignocellulosic biomass is mainly composed of cellulose, hemicellulose and lignin along with minor fractions of proteins, pectin, extractives and ash (Kumar et al., 2009). First commercial plants for a biochemical conversion of lignocellulosic biomass are in operation with annual capacities up to 75,000 tons. These process mainly biomass with a low lignin content, while biorefineries based on agricultural residues are still under development (Brown et al., 2015). In order to accelerate the development of these biorefineries, the case studies in this thesis focus on the processing of lignocellulosic biomass residues.

2.2 Process development

Process development is divided into four development stages, namely the product and pathway selection, the conceptual process design, detailed engineering and construction (Biegler et al., 1997). Since the process knowledge and therefore model complexity at the first stage (product and pathway selection) is low, the uncertainty in the resulting design is high. At the following stages, the model complexity and accuracy raises whereas the number of alternatives is stepwise reduced. A tedious detail engineering is only conducted for the most promising process concept. Therefore, the decisions at the first two stages (product and pathway selection, conceptual process design) have a high impact on the final process efficiency and are responsible of up to 80% of the final cost (Biegler et al., 1997). Therefore, these decisions need to be addressed systematically (Mitsos et al., 2018).

For a given product and pathway, the process development directly starts with the conceptual process design. The most common approaches are the onion model by Smith and Linnhoff (1988) or the hierarchical design procedure proposed by Douglas (1985), which decompose the procedure into subproblems.

2.2 Process development

The subproblems are the decision on a batch versus a continuous production, the development of an input-output structure followed by a recycle structure and a separation system structure. Finally, the heat integration potential and a heat exchanger network are examined. At each level, different process alternatives are conceivable. If the decision on these design questions is postponed, the process development procedure results in a tree of alternatives.

If the decision on a reaction pathway or even on the desired product is a degree of freedom as well, the process development soon results in a combinatorial explosion of alternatives. Even though, commercial simulation software tools support the design procedure, a conceptual design for all resulting alternatives is costly and tedious. In order to setup a process simulation, design decisions need to be predefined. Due to a high number of possibilities for the design specifications, the development and identification of globally optimal processes is challenging. Thus, a comparison of different processes always depends on user-specified design decisions. In addition, commercial simulation software often lack the required robustness for a comprehensive analysis of many process options. If only few setups are selected for a conceptual design study, e.g., based on heuristics, there is a risk of missing the most promising alternatives.

Therefore, all possibilities need to be considered simultaneously in order to identify the best performing pathway (Westerberg, 2004). For this purpose, a superstructure is used in which the different process alternatives (configurations, technologies) are summarized. Binary variables indicate the absence or presence of a certain alternative. The resulting optimization problems are therefore often mixed-integer nonlinear programs (MINLP) and the objective of the optimization is to identify the most promising alternative (Grossmann, 1990; Biegler et al., 1997)

In literature, superstructures are well established, for instance, for the systematic identification of optimal mass and heat exchanger networks and for the structural optimization of flowsheets (e.g., El-Halwagi and Manousiouthakis, 1989; Srinivas and El-Halwagi, 1994; Papalexandri and Pistikopoulos, 1996; Biegler et al., 1997). In order to keep the optimization problem at a manageable level, the superstructure is either applied for a subproblem only, like a heat exchanger network, or the model complexity of the unit operations is reduced to a simple input-output model.

2.3 Process screening approaches for biorefineries

Within the context of biorefineries, a high number of reactions, reaction pathways (e.g., Huber et al., 2006; Alonso et al., 2010; Serrano-Ruiz et al., 2010; Serrano-Ruiz and Dumesic, 2011; Besson et al., 2014; Climent et al., 2014) and also several processing concepts (e.g., Hayes et al., 2006; Humbird et al., 2012) have been proposed and published in literature. If all of the different process opportunities are considered, a combinatorial explosion is caused. Thus, efficient decision support for the identification of the most promising processes is mandatory (Mitsos et al., 2018).

The selection of optimal processing pathways is of particular interest, as it represents the highest contribution to the value chain. Hence, giving guidance to future research directions by reducing the large amount of alternative reaction and processing pathways down to the most promising pathways, an early design stage screening methodology is required. Several approaches and methods tackling this challenging problem exist in literature, which mainly rely on a superstructure formulation. Within the context of biorefineries, Sammons et al. (2008) proposed a general workflow for the setup of a superstructure formulation. It starts with the setup of simplified data-driven models for the process performance of selected technologies, based on either extensive literature searches or simulation studies. In a second step, a superstructure is formulated, which is used to evaluate the process performance of biorefinery concepts.

Based on this general procedure, several groups propose superstructure formulations for various fuels and other products (e.g., Bao et al., 2011; Santibanez-Aguilar et al., 2011; Rangarajan et al., 2014a; Cheali et al., 2014; Kelloway and Daoutidis, 2014; Zondervan et al., 2011; Celebi et al., 2017). Santibanez-Aguilar et al. (2011) present a superstructure for biodiesel, ethanol and hydrogen using multi-objective optimization to simultaneously minimize the economic, environmental and social impact. Cheali et al. (2014) analyze the thermo- as well as biochemical production based on similar data-driven models as main input. In contrast, Kelloway and Daoutidis (2014) consider in their superstructure formulation yield or conversion constraints for biochemical conversion and equilibrium and kinetic models in case of thermochemical conversion. Zondervan et al. (2011) discuss a multi-product formulation of biorefinery systems. Extending the aforementioned models by considering the energy demand of selected unit operations and technologies, several groups propose extended superstructure formulations analyzing in addition, e.g., the global warming potential or the overall process energy demand (Kim et al., 2013; Wang et al., 2013; Rizwan et al., 2015; Quaglia et al., 2015; Garcia and You, 2015b; Gargalo et al., 2016a,b).

2.3 Process screening approaches for biorefineries

These formulations vary in their complexity as these data-driven models are either generally valid for a process or are specific for particular unit operations (Garcia and You, 2015b). An additional heat integration is performed by Kokossis et al. (2015), Tock and Maréchal (2012), Niziolek et al. (2015), Kong and Shah (2016) and Celebi et al. (2017). In Martín and Grossmann (2013) water integration is considered. Further work exist addressing the uncertainty in the design of biorefineries (Tang et al., 2013).

All of these superstructure-based methods are capable of evaluating existing process schemes and connections thereof, but have several restrictions in common. The data-driven input-output models are highly specific for the respective process or unit operation, inlet composition and operating conditions. Hence, feasibility or accuracy of the unit operations cannot be generalized as, for instance, the energy demand of a separation strongly relies on the thermal state of the inlet stream. The approaches further require either an extensive literature research for the split and conversion factors as well as for the cost and energy demands or a high number of simulation studies often lacking robustness, rendering the overall procedure tedious. Kokossis et al. (2015) report preparation times longer than a year, postponing early stage design decisions.

In order to reduce the time and effort for data collection and ensure feasible connections between individual processing steps, ontology-based data storage is proposed in literature. Besides data storage, the Super-O implementation of Bertran et al. (2017) further includes consistency checks and a subsequent automated network formulation and solution. Furthermore, feasible connections from process analysis to supply chain and product selection are possible. An alternative attempt tries to reduce the data collection time by gathering the literature knowledge available at a web-based interactive platform (Siougkrou and Kokossis, 2016a,b). While both approaches reduce the effort to setup the required superstructures and efficiently store available data for future reuse, the data is still restricted to existing process designs from literature or simulation studies only. Therefore, the available approaches are not capable of efficiently evaluating novel pathways and products.

A first step towards a systematic and fast evaluation of novel reaction pathways is the RNFA proposed by Voll and Marquardt (2012a). Since the proposed methodology in this thesis builds upon the RNFA, the RNFA concept, general assumptions, model formulation as well as current shortcomings are described in Section 2.4. Note, that the same mathematical basis, i.e., a superstructure formulation, for all screening approaches is used, such that in principle any alternative screening method could have been used as starting point for this thesis as well. Therefore, the reasons to choose the RNFA are further elaborated.

2 Screening approaches for biorefinery processes

The target of this thesis is to bridge the gap between the selection of reaction pathways and conceptual design by introducing a screening method for process pathways. This results in a biorefinery process framework for a systematic analysis of novel reaction (step 1) and process pathways (step 2), which gradually reduces the number of alternatives for the subsequent conceptual design (step 3). The framework is inspired by the process synthesis framework of Kraemer et al. (2009) and Skiborowski et al. (2013). The process synthesis framework decomposes the process design for a single separation task and is therefore based on sequential model refinement to obtain cost-optimal processes. The biorefinery process framework is visualized in Figure 2.1.

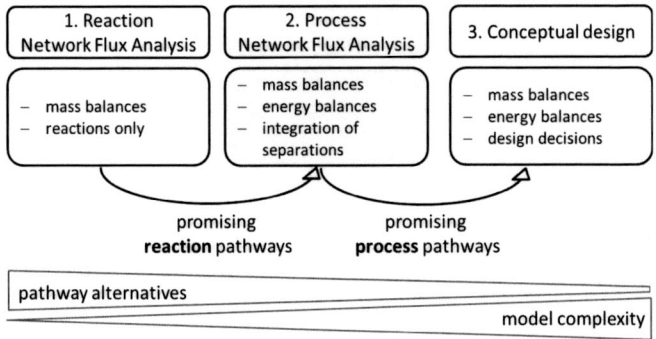

Figure 2.1: Biorefinery process framework for the assessment of novel process concepts. In a first step, the mass balances of various reaction pathways are analyzed by means of Reaction Network Flux Analysis. In a second step, Process Network Flux Analysis is applied, which incorporates the choice of separations and, thus, energy balances. Finally, for the most promising process pathways, a conceptual design is conducted, in which design decisions are made.

Whereas, the RNFA considers reaction pathways only, the novel screening method Process Network Flux Analysis extends the analysis opportunities to an integration of separation steps. In the conceptual design, design decisions like the column height, diameter or feed stage are then selected. Therefore, the model is sequentially refined and problem complexity is increased. This procedure reduces the effort in the conceptual process design phase, as it directly starts from the most promising pathways.

The alternative screening methods rely on data input obtained in the conceptual design phase and therefore, do not support the proposed framework for the assessment of novel processes. Even though, these methods could have been adapted to incorporate the assessment of novel processes in a similar manner, it is wise to use the RNFA as starting point in this thesis, such that the input format and model are similar for the first two steps of the framework to ensure an easy data and knowledge transfer from the reaction to the process screening.

2.4 Reaction Network Flux Analysis

RNFA aims at the evaluation and benchmarking of existing and novel reaction pathways. The methodology relies on laboratory reaction data. Therefore, novel reactions and reaction pathways can be evaluated rapidly, which enables an early detection of bottlenecks. Thus, RNFA accelerates the development of viable and sustainable processes (Voll and Marquardt, 2012a).

The RNFA has been successfully applied for various case studies, such as in the production analysis of fuels (Voll and Marquardt, 2012a,b), medicines (Gantner, 2015) or bio-polymers (Zhang et al., 2017; del Rio-Chanona et al., 2018). Similar to metabolic flux analysis (Schilling and Palsson, 1998), the basis for RNFA is always a reaction network containing various starting materials (e.g., different biomass types and compositions), intermediates as well as products (e.g., fuels, value-added chemicals), linked and connected by the respective reactions. In Figure 2.2 a network scheme is shown.

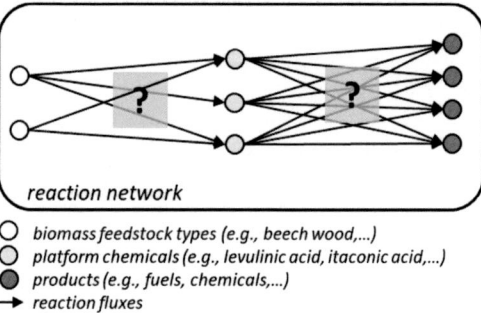

Figure 2.2: Reaction network scheme. Similar to metabolic networks, arcs visualize reactions and nodes represent molecules.

Similar to metabolic networks, nodes represent the components and arcs the reactions according to Voll and Marquardt (2012a). The systematic setup of these reaction networks is described in the next section.

2.4.1 Network construction and input data

The reaction networks might either be generated based on literature search (Voll and Marquardt, 2012a) or automatically by means of reaction rules (e.g., Rangarajan et al., 2012, 2014b; Bertók and Fan, 2013). In automatically generated networks, often the reagents as well as the desired product(s) need to be specified. Based on reaction rules and a maximal number of conversion steps, the reagents and products are interconnected. In principle, both directions namely the forward direction converting the reagent into the product as well as backwards from the product to the reagent, or even a combinations thereof can be utilized to derive these networks (Pham and El-Halwagi, 2012). Instead of defining one or multiple desired products, product specifications can be used to characterize the product performance, while cutting off those product molecules and their respective pathways, which do not satisfy the product specifications. Even product mixtures can be targeted, if the product properties can be addressed, e.g., by linear mixing rules (Victoria Villeda, 2016). The underlying reaction rules are formalisms of well-known chemical reaction classes, like the hydrogenation of alkenes to alkanes. Independent of the algorithm and software chosen, these automatically generated networks all have the advantage of proposing and suggesting novel reaction pathways, which otherwise might not be considered. Hence alternative pathways are revealed opening up new alternatives (Victoria Villeda, 2016).

As the network size correlates with the number of underlying reaction rules, these networks are restricted to the range of applied reaction rules. While a reaction might be theoretically possible, in reality the reaction performance depends on the catalyst performance, solvent and reaction conditions. Therefore, feasibility of the automatically derived reactions cannot be guaranteed, as for the majority of chemical and biological reactions a tailored catalyst is mandatory, which lowers the activation energy of a reaction and, thus, enables the reaction in the first place.

Manually assembled reaction networks are time-consuming as they require an extensive literature search. However, using online databases like Sci-Finder (American Chemical Society, 2015) or Reaxys (Elsevier Information Systems GmbH, 2015) and the forward and backward approach of Pham and El-Halwagi (2012), even large networks can be set up fairly easily in a short time. This procedure is simplified by tools like the synthesis plan of Reaxys (Elsevier Information Systems GmbH, 2015).

Herein, a starting component (reagent, product) needs to be specified to automatically develop a synthesis tree. Similar to the above mentioned network generators, the structure of the synthesis tree can be influenced by restricting, e.g, the number of steps or the type of co-reagents.

The synthesis plan can then be used to set up large reaction networks and download key reaction parameters like the reaction yield, and conditions. In the master thesis of Gantner (2015), an automatic assembly of a network using the Reaxys database (Elsevier Information Systems GmbH, 2015) has been conducted. Based on a maximal pathway step length of five, a large network with more than 16 million reactions has been successfully set up. However, the analysis of the network has been challenging as in many cases the reaction entries were incorrect, important information about the reaction was (partially) missing or false and reaction stoichiometry was often not fully stored in the database (Gantner, 2015). While a data verification is intractable for these very large networks, for smaller networks data validation is not a big effort. Hence, in this thesis, manually assembled reaction networks are utilized to ensure the feasibility of a reaction. The database information is checked manually. With a further improvement of the database, even larger networks might be generated and set up in the future.

Along with information about the reaction like the reaction stoichiometry and yield, the RNFA requires pure component properties, which are easily accessible via literature search or group-contribution methods (e.g., Joback and Reid, 1987; Gani et al., 1997; Marrero and Gani, 2001).

2.4.2 Model formulation

In the following, the RNFA model is briefly presented with an emphasis on the main equations to setup the optimization problem addressing the economic efficiency and sustainability of reaction pathways. For a detailed description as well as information on the model development and derivation of equations, refer to Voll (2014). Basis of the RNFA is a molar flux balance, balancing the sources (reactants) of a network as well as the sinks (products) as expressed in Equation 2.1,

$$\boldsymbol{A} \cdot \boldsymbol{f} = \boldsymbol{b}, \quad \boldsymbol{A} = \begin{bmatrix} \boldsymbol{A}_1 & \boldsymbol{A}_2 \end{bmatrix}, \quad \boldsymbol{f} = \begin{bmatrix} \boldsymbol{f}_1 \\ \boldsymbol{f}_2 \end{bmatrix}, \qquad (2.1)$$

where $\boldsymbol{A}$ denotes the stoichiometric matrix, which contains the stoichiometric coefficients of all the components in a reaction. The vector $\boldsymbol{f}$ represents normalized reaction rates as fluxes and $\boldsymbol{b}$ is the product vector (Voll and Marquardt, 2012a).

The stoichiometric matrix consist of a sub-matrix A_1 and the vector $\mathbf{f}$ has a sub-vector f_1, in which components are only supplied to the network and no reaction takes place. Consequently, A_1 is named supply matrix and f_1 supply fluxes. The sub-matrix A_2 covers reactions and is therefore referred to as reaction matrix and the fluxes f_2 as reactive fluxes (Voll and Marquardt, 2012a).

The fluxes are constrained by the reaction yield. In literature, most publications present reaction yields based on the result from laboratory experiments as the only published parameter (e.g., Dumesic et al., 2010), not specifying in detail the conversion or the selectivity of a reaction. Hence, both, the RNFA and the extension to the PNFA, are based on the yield as input parameter as a clear specification and differentiation between conversion and selectivity is often not given and therefore not feasible. The following inequality constraint presents the flux limitation of reaction j by a reaction yield Y, taking into account a key reactant kr and main product mp,

$$f_j \leq Y_{mp,j} \frac{\sum\limits_{in} \nu_{kr,in} f_{in} + \sum\limits_{out, out \neq j} \nu_{kr,out} f_{out}}{\nu_{mp,j}}. \tag{2.2}$$

Herein, the stoichiometric coefficients ν as well as the sum of all incoming f_{in} and outgoing fluxes f_{out} are considered (Voll, 2014).

One basic RNFA assumption is a conversion limitation (Voll and Marquardt, 2012a,b), which is briefly explained in the following. Reactions with a yield lower than the theoretical yield, are either limited by a maximal conversion (e.g., equilibrium reactions), by a maximal selectivity or by a combination of conversion and selectivity. Hence, the concepts of a conversion or selectivity limitation represent extreme cases. If these extreme cases are compared, the selectivity limitation represents the worst and the conversion limitation the best case, as in the first case waste is produced and needs to be disposed and in the second case unconverted reagents might be recycled or react in any subsequent or parallel reaction. A (known) reaction selectivity is easily inserted into the stoichiometric matrix $\mathbf{A}$ which converts the yield constraint into a conversion constraint. Thus, conversion limitation is assumed for both the RNFA and PNFA representing a best-case scenario.

The main biorefinery feedstock analyzed in the RNFA and this thesis, is lignocellulosic biomass, which is a composite material, consisting mainly of cellulose, hemicellulose and lignin. For simplicity reasons only these major biomass fractions are considered in the network analysis (Voll, 2014). Depending on the type, region, season and weather conditions, the biomass composition varies significantly, such that a general formula for the biomass composition does not exist. Instead, only wide ranges of the biomass fractions are given in literature.

2.4 Reaction Network Flux Analysis

Therefore, lower bounds x_{low} and upper bounds x_{up} for the three main fractions are introduced into the optimization problem,

$$x_{low,C_6H_{10}O_5} \leq x_{C_6H_{10}O_5} \leq x_{up,C_6H_{10}O_5}, \qquad (2.3)$$

$$x_{low,C_5H_{10}O_5} \leq x_{C_5H_{10}O_5} \leq x_{up,C_5H_{10}O_5}, \qquad (2.4)$$

$$x_{low,C_{10}H_{12}O_3} \leq x_{C_{10}H_{12}O_3} \leq x_{up,C_{10}H_{12}O_3}, \qquad (2.5)$$

with a representative molecular structure for cellulose $C_6H_{10}O_5$, hemicellulose $C_5H_{10}O_5$ and lignin $C_{10}H_{12}O_3$, as proposed by Voll (2014). According to Mergner et al. (2013), the hemicellulose fraction is assumed to be fully hydrolyzed after pretreatment. A varying biomass composition enables the solver to choose the optimal biomass composition. In case of a known specific composition, the molar fractions x can be fixed. The closing condition,

$$x_{C_6H_{10}O_5} + x_{C_5H_{10}O_5} + x_{C_{10}H_{12}O_3} = 1, \qquad (2.6)$$

implies that the molar fractions of the main components need to add up to one (Voll, 2014).

As outlined in Section 2.1, the biochemical conversion involves a biomass pretreatment as a first step. However, a simplified and correct description of the various biomass pretreatment alternatives is not yet available, therefore the biomass split into its constituents is realized by simple reaction fluxes (Voll, 2014). As a similar biomass pretreatment is required for all pathways, this assumption does not influence a product ranking.

Based on the flux balance, yield constraint and biomass composition, a first insight into the economic efficiency and sustainability of biorefinery reaction pathways is achieved. In the work of Voll and Marquardt (2012a), the economic efficiency is expressed as total annualized costs (TAC),

$$TAC = \frac{IC \cdot ir}{1 - (1 + ir)^{-n}} + \sum_{i=1}^{s} f_{s,i} P_i, \qquad (2.7)$$

which comprises annualized investment costs IC and raw material costs. The latter can be easily determined based on the supply fluxes f_s of raw materials and the corresponding prices P. The IC are annualized based on an interest rate ir and a plant run time n.

These IC are estimated using the heuristic approach of Lange (2001), which correlates the IC to the enthalpy of combustion loss of processes. Thus, the plant capacity or pathway complexity in terms of the number of required apparatus, are not included.

2 Screening approaches for biorefinery processes

A detailed discussion thereof as well as an overview of available IC approaches applicable at early stage is conducted later in Section 3.2.2.1. As a second objective function, the sustainability of the pathways is addressed. Various metrics are conceivable like an external hydrogen consumption, carbon efficiency or a lumped environmental impact factor EI according to Uhlman and Saling (2010). The EI captures the loss of the enthalpy of combustion (EC), the resource consumption (RC), contributions from renewable stoichiometric emissions (Em) and the toxicity potential of the final product (ToxP). The individual factors are normalized on their maximal value of the case study and summed up to an overall environmental impact,

$$EI = \frac{EC}{\|EC\|_{maximal}} + \frac{RC}{\|RC\|_{maximal}} + \frac{Em}{\|Em\|_{maximal}} + \frac{ToxP}{\|ToxP\|_{maximal}}. \quad (2.8)$$

The overall EI is then normalized once more on the worst case result of the current case study $\|EI\|_{maximal}$ to receive values between zero and one (Voll, 2014). This iterative procedure for the determination of the worst case scenario and hence the normalization factors is tedious. In addition, the normalization factors are very specific for a single case study, such that comparability of different case studies is not possible. Therefore, an adapted form of the EI is applied later in Section 3.2.3.

With the introduction of a heating value equivalent α as design target, the optimization problem can then be formulated as

$$\min_{f,b} \left\{ \begin{array}{c} TAC \\ EI \end{array} \right\}$$
$$\text{s.t.} \; \boldsymbol{A} \cdot \boldsymbol{f} = \boldsymbol{b},$$
$$yield\ constraint\ (Eq.\ 2.2),$$
$$biomass\ composition\ (Eqn.\ 2.3 - 2.6),$$
$$TAC\ calculation\ (Eqn.\ 2.7), \quad (2.9)$$
$$EI\ calculation\ (Eqn.\ 2.8),$$
$$\sum_{i=1}^{N_{products}} b_{product,i} \Delta H_{comb,product} = \alpha,$$
$$\boldsymbol{f}, \boldsymbol{b} \geq 0,$$

which is a nonlinear programming problem. The nonlinearity arises from the exponent of the investment cost function. All required model parameters together with their associated references are given in Appendix C.1.

2.4.3 Shortcomings

In order to identify potential improvements and requirements for a further RNFA development, the shortcomings are briefly summarized. The formulation of RNFA relying on yield data only, is both a blessing and a curse. While it allows for a reaction pathway screening based on mole balances, it clearly neglects solvent utilization and the separation. However, separation is still responsible for the major share of the overall energy demand (Sholl and Lively, 2016) and, thus, production cost, such that a simultaneous consideration at an early stage is preferred.

To benefit from the additional knowledge gain by the integration of separations, the economic and sustainability evaluation criteria need to be reviewed once more to identify the most relevant and informative metrics for the analysis of reaction and processing pathways. The TAC for example, do not cover utility or waste disposal cost. The IC estimation according to Lange (2001), which is based on EC, is counterintuitive as the correlation calculates high IC for processes with a large energy demand. According to this correlation, processes using heat integration and therefore less external energy, reduce the IC. However, in reality, additional heat exchanger or heat pumps are required, which increase the IC. The application of the EI is difficult as well, as it requires a case-study specific normalization on the worst case result. Therefore, comparability of different case studies is impossible. Hence, alternative criteria and their applicability are discussed in Section 3.2.3.

RNFA is restricted to the analysis of reactions only. The effect of biomass transportation on the cost is not considered, although an efficient supply chain is mandatory for economically viable concepts. Market and price uncertainties also play a major role regarding profitability analyses as well as product portfolio selection. Even though, a first step towards an integration of market modelling into the RNFA has been proposed (Sorda and Madlener, 2012; Voll, 2014), it requires knowledge of many and very specific parameters often not available at early stage. In the following, existing work on the consideration of separations at early stage is summarized before the literature on value chain analysis, i.e., biomass supply chain and market modelling is discussed.

2.5 Consideration of separation steps

Several empirical approaches are available to address the separation effort at early stage, e.g., categorizing the separation complexity by a weighted sum of empirical indicators considering, e.g., the presence of solids or azeotropes (Moncada et al., 2015).

In terms of energy demand, correlations for various unit operations are known in literature (Hermann and Patel, 2007; Albrecht et al., 2010). However, these equations are either based on expert knowledge or fitted to historical data hence might under- or overestimate the energy demand strongly. In addition, empirical correlations for non-ideal and especially azeotropic mixtures do not exist.

Recently, Wu et al. (2016) proposed a superstructure-based framework for bio-separations, which relies on heuristics for unit operations like adsorption, crystallization and precipitation. An explanation on the determination of the separation factors for the heuristic models is not given. In addition, simplified shortcut models for distillation, i.e., the Underwood-Fenske equations for ideal mixtures are applied by Wu et al. (2016) and Kong et al. (2016) to determine ideal distillation sequences.

Simplifications like the assumption of ideal mixtures for distillation might be misleading, since bio-molecules often exhibit non-idealities like azeotropes. Hence, for an accurate performance evaluation the models should rely on rigorous thermodynamics.

In order to accurately describe non-ideal vapor-liquid equilibria either an activity coefficient model, i.e. a so-called G^E-model, for describing the non-ideality of the liquid phase, or an equation of state (EOS) is required, which could be used to describe non-idealities in the liquid and the vapor phase.

A general challenge for modeling these system is however, that for bio-based components and mixtures thereof, property parameters are still not available for most G^E-models and EOS. Furthermore, an accurate parameter estimation is challenging, as for most bio-based components sufficient experimental data is still missing. Therefore, property estimation based on group-contribution methods or quantum-mechanical calculations are the only available options. However, common group-contribution methods like UNIFAC (Wittig et al., 2003) or modified UNIFAC (Jakob et al., 2006) often lack some of the necessary parameters for important molecular groups, as, e.g., the furan ring, which is typical in bio-based molecular building blocks (Dortmund Data Bank Software & Separation Technology GmbH, 2016).

Advanced EOS based on group contributions like the SAFT γ-Mie model (Papaioannou et al., 2011; Dufal et al., 2014), offer the advantage of a thermodynamically consistent computation of an extended range of thermodynamic properties, but at current state have an even more limited range of available group contribution parameters as the classical group contribution G^E-models UNIFAC and modified UNIFAC.

Quantum-mechanical methods like the *Conductor-like Screening Model for Realistic Solvents* (COSMO-RS) are an alternative to predict the property parameters, as long as the required group contribution parameters for the other models are still missing.

Since the Non-Random Two Liquid (NRTL) model is a prominent example for the activity coefficient models and parameter sets for a large number of mixtures exist in literature, it is applied in this thesis. The NRTL parameters are taken from literature databases whenever available or are predicted using COSMO-RS (Klamt, 2005).

2.6 Value chain analysis

In order to obtain viable and sustainable processes, process screening needs to be integrated with biomass supply chain design and market modelling as early as possible. This allows the analysis of full value chains from biomass transportation via processing to product portfolio selection. A scheme of a value chain analysis is illustrated in Figure 2.3.

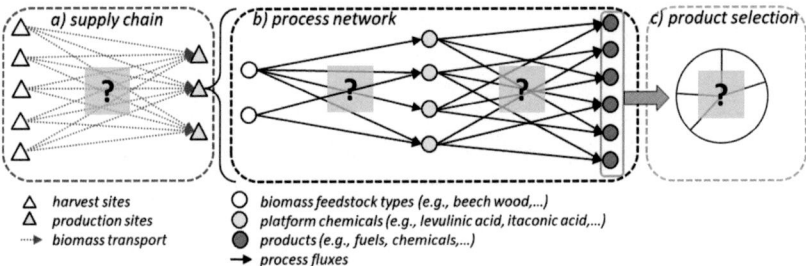

Figure 2.3: Value chain scheme for biomass supply chain a), coupled to process network analysis b) and product-portfolio selection c). Herein, biomass transportation connects the various biomass harvesting to the potential plant sites. The resulting biomass supply chain is specific for each production site. For every site, a biomass processing concept is determined as result of an optimal process network analysis b). For a multi-product biorefinery, the process network is further constrained by the determination of an optimal product-portfolio selection c). The selection is based on the maximization of revenues and calculates an optimal product massflow distribution.

In Figure 2.3 a), a biomass supply chain is visualized connecting the different harvest to the various production sites. The dotted arrows indicate the biomass transport from the harvest to the production site. For every production site an optimal process pathway(s) is (are) selected for single or multi-product biorefineries (Figure 2.3 b)).

Herein, lignocellulosic biomass is converted via its constituents to platform chemicals and final products. Figure 2.3 c) presents the product-portfolio selection depending on market and price developments.

2.6.1 Biomass supply chain

Compared to lower heating values of 42 $\frac{MJ}{kg}$ for crude oil (Huber et al., 2006; Serrano-Ruiz and Dumesic, 2011) and 40-44 $\frac{MJ}{kg}$ for petroleum fuels (Demirbas, 2011), the heating value of biomass is low, approximately 15-20 $\frac{MJ}{kg}$ depending on the biomass type and composition (Huber et al., 2006; Serrano-Ruiz and Dumesic, 2011). In addition, biomass is spread over large areas, such that transportation can contribute significantly to the production cost (Demirbas, 2011; Cambero and Sowlati, 2014). In fact, Angus-Hankin et al. (1995), Kumar et al. (2006) and Ekşioğlu et al. (2009) determined that transportation of forestry biomass accounts for 20-40% of the fuel cost. This is one of the main reasons preventing lignocellulosic ethanol plants from commercialization (Ekşioğlu et al., 2009). Thus, transportation needs to be integrated into the process analysis at early stage for the identification of viable processes.

Smaller plant sizes compared to the fossil-feedstock based industry and, thus, distributed production are targeted to reduce transportation costs of biorefineries (Serrano-Ruiz and Dumesic, 2011). As a consequence of the economy of scale, the relative investment cost contribution is reduced for larger plant sizes, thus, targeting centralized production. These contradicting effects can be addressed using the overall production costs or greenhouse gas emissions to determine optimal plant capacities and locations (You and Wang, 2011). Examples are given by You and Wang (2011) and Lara and Grossmann (2016), who describe optimal combinations of centralized and distributed production by simultaneous supply chain and capacity analysis. This illustrates that an integrated process and supply chain design at early stage is required to screen for the most viable solutions (Serrano-Ruiz and Dumesic, 2011; You and Wang, 2011).

In order to analyze the effect of transportation on processes viability, process screening needs to be integrated with supply chain design. For this purpose, a profound list of supply chain models exist in literature (cf. Shah (2005); Papageorgiou (2009); Garcia and You (2015b) for general reviews and Cambero and Sowlati (2014) as well as Ghaderi et al. (2016) for forest biomass supply chain) varying in the number of analyzed aspects and, thus, in their complexity. The availability of different biomass compositions (You and Wang, 2011; Sharifzadeh et al., 2015) and their seasonality (Ekşioğlu et al., 2009; You and Wang, 2011; Čuček et al., 2014; Avraamidou and Pistikopoulos, 2017) strongly influences the determination of an optimal supply chain.

2.6 Value chain analysis

However, detailed biomass availability and distribution data is scarce. The biomass transportation includes different transportation modes (rail, ship, truck) (Schwaderer, 2012) and biomass storage (Ekşioğlu et al., 2009; Ng and Maravelias, 2017a,b; Ng et al., 2018). The identification of optimal plant locations (You and Wang, 2011; Čuček et al., 2014; Ng and Maravelias, 2017a,b) depends on the decision regarding distributed or centralized production or a combination thereof to obtain more flexibility (You and Wang, 2011; Sharifzadeh et al., 2015; Lara and Grossmann, 2016).

Since data is scarce, a pragmatic supply chain is chosen in a first step, considering biomass availability and transportation to identify optimal plant locations and analyze process viability for a distributed production. Like the existing work on supply chains, an economic (Kim et al., 2011a,b; Schwaderer, 2012; Elia et al., 2013; Marvin et al., 2013b; Čuček et al., 2014; Sharifzadeh et al., 2015) and environmental (You and Wang, 2011) objective is utilized in the optimization. Thus, the effect of social aspects as proposed by You et al. (2012) and Cambero and Sowlati (2014) is not included. Finally, the analysis is complemented by a sensitivity analysis to address the effect of uncertainties as proposed by Kim et al. (2011b), Sharifzadeh et al. (2015) and Gao and You (2017).

All of the aforementioned models are connected either to a biochemical ethanol production (Ekşioğlu et al., 2009; Huang et al., 2010; You et al., 2012; Marvin et al., 2013b; Ng et al., 2018), a thermochemical conversion plant (Kim et al., 2011a; You and Wang, 2011; Elia et al., 2013; Sharifzadeh et al., 2015) or a combination thereof (Čuček et al., 2014). Thus, the number of products and hence processing alternatives is limited. Furthermore, the products are already preselected. In this thesis, supply chain design is utilized to determine biomass transportation cost and global warming. Thereby, the influence of biomass transportation on process viability is analyzed. Furthermore, the number of alternatives is increased to include novel process concepts.

2.6.2 Market models

Large scale biorefinery plants for bio-fuel production are difficult to realize due to the high transportation costs. Small scale plants are not competitive as these are missing the benefit from economy of scale. Thus, a profitable bio-fuel production is still ambitious.

In order to obtain profitability, the co-production and sale of value-added chemicals is an attractive opportunity (Tsakalova et al., 2014; Pyrgakis and Kokossis, 2016; Celebi et al., 2017). For this purpose, the influence of value-added products on the profitability, obtainable prices and market sizes need to be considered simultaneously.

This is especially important at early stage, as the product portfolio selection depends on these market parameters and influences the viability of a biorefinery (Sorda and Madlener, 2012; Voll, 2014). Thus, a profound model covering the biomass supply chain, processing and product alternatives as well as market developments is required.

Price and market uncertainties need to be considered for both the biomass feedstock as well as for the final products (Mansoornejad et al., 2013; Shabani et al., 2013; Kokossis et al., 2015; Giuliano et al., 2016; Black et al., 2016), which are subject to supply and demand principles. Thorough market models and resilient forecasts for bio-based chemicals are not available in the open literature and publication is often restricted by companies (Mansoornejad et al., 2013). Economic forecasts are particularly challenging in the non-established field of biorefinery markets. Therefore, market models are often simplified and can only be used for a first insight.

For this purpose, sensitivity analyses are the most simple approach to consider price uncertainties and analyze the effect of fluctuating prices on the product selection (Mansoornejad et al., 2010; Kim et al., 2011a). Alternative approaches derive future scenarios for market uncertainties (Sepiacci et al., 2017), correlate prices of bio-products to an oil-price model or adjust them based on fossil replacement or substitution products (Sharma et al., 2013; Geraili and Romagnoli, 2015). This is restricted to fossil replacement products, such as bio-polyethylene or substitution products exhibiting the same functionality as the fossil counterpart. These scenarios or oil-price correlation models have a high degree of nonlinearity and are based on a large number of assumptions, like estimates of the replaced fossil processes or feedstock margins (Sharma et al., 2013). Furthermore, the models omit the influence of the market size on attainable product prices. To prevent an excess production, a constraint can be added to guarantee a certain customer demand (Sharifzadeh et al., 2015) or to restrict the maximum product demand (Kim et al., 2011a; Cambero et al., 2016). While an excess production is prohibited, the dependency of prices on the market exploitation is not depicted. Existing biomass market models by Sorda and Madlener (2012) and Voll (2014) take these dependencies into account, but rely on a high number of uncertain parameters, like price elasticities, which vary between different biomass types by more than one order of magnitude and are region specific (Simangunsong and Buongiorno, 2001). Due to large biorefinery market uncertainties and a significant increase in nonlinearity, complex market models are not suitable for an early stage design. Instead, a pragmatic approach is targeted in this thesis (cf. Section 3.4).

2.7 Uncertainties at early design stage

All screening approaches as well as the biomass supply chain and market modeling rely on data and model assumptions, which are subject to uncertainty. Uncertainties from model assumptions arise from simplifications like the biomass split into its constituents or a rough IC estimation. These simplifications are mandatory, as detailed knowledge is not available at early stage. The influence of these assumptions on the accuracy of the results is discussed in detail in Chapter 4.

Uncertainties of the PNFA resulting from the required input data need to be addressed systematically. In particular, lab-scale yields for reactions and shortcut results for separation are typically used, though they cannot directly be transferred to an industrial-scale process. If these uncertainties are not explicitly addressed in process design (e.g., Halemane and Grossmann, 1983; Pistikopoulos, 1995; Sahinidis, 2004; Steimel and Engell, 2016), the assessment resulting from the RNFA or PNFA will not be fully satisfactory. In order to integrate the influence of uncertainties into the analysis, different approaches exist. Robust optimization methods as proposed by Sy et al. (2018) leads to robust, but worst-case statements and, thus, to conservative designs (Mitsos et al., 2018). Another opportunity is to couple the evaluation with a sensitivity analysis which is defined as the dependency of the model results on model structures and parameters employed (Saltelli et al., 2004).

A sensitivity analysis in the context of pathway screening approaches addresses two questions: (i) Is the ranking of different pathways still reliable for a given level of uncertainty? (ii) Which model parameters are most influential when determining the rank order of pathways? If the most critical parameters are identified this way, the associated uncertainty can be reduced in further experimental or theoretical investigations to increase the accuracy of the screening results. Different approaches to sensitivity analysis exist in literature, including response surface methods, differential analysis and Monte Carlo analysis (Iman and Helton, 1988). A sampling-based Monte Carlo approach is capable of predicting the worst-case scenario which may result in case of significant variations in multiple parameters. Here, the one-at-a-time analysis (OAT) (Saltelli et al., 2007) is introduced into the PNFA, where only variations in a single parameter are performed while the residual parameters are fixed to their nominal value. In this way, the most sensitive parameters and those which only have negligible effect on the results can be identified.

Existing work in the field of biorefineries considers uncertainties of yield (Rizwan et al., 2015), economic (Quaglia et al., 2013; Tock and Maréchal, 2015) as well as feedstock supply and market parameters (Kim et al., 2011a; Tang et al., 2013).

Accordingly, the parameters varied in this thesis can roughly be classified into five groups, namely property and process data, profitability, sustainability and supply chain parameters.

Uncertain property data include pure component as well as mixture properties. Pure-component properties, like the enthalpies of formation and combustion, are crucial for both the RNFA and the PNFA, whereas mixture properties are only relevant for the PNFA. Although uncertainties in the mixture parameters are expected, it is difficult to obtain a thermodynamically consistent model if only a single parameter is altered. Thus, the effect of uncertainties in the mixture properties is only addressed by varying the resulting energy or auxiliary demand of separations.

The process data consists of reaction and separation data, with the latter only being relevant for the PNFA. In case of reaction data, the reaction stoichiometry is assumed to be correct, such that uncertainty in the yield coefficients is analyzed. The uncertainty in the reaction yield results from scale-up from lab- to plant-scale and often also from transferring the batch to a continuous production. The separation parameters cover the separation split as well as the energy and auxiliary requirement for a specific separation.

Further parametric uncertainty is related to the market and price parameters as well as to the sustainability parameters. In the supply chain design, the available biomass capacity at a single origin and the water content are subject to uncertainties. To determine uncertainties related to process evaluation, a case-study specific list of parameters for a OAT analysis is described in Section 5.1.5. In order to address additional uncertainties arising from supply chain parameters, an extended list of parameters is analyzed in Section 5.1.7.3, which includes supply chain parameters. In the following, requirements for a novel process screening method are summarized.

2.8 Requirements for a novel screening method

To guide future research by quickly evaluating process concepts, a novel screening method is mandatory which is able to assess existing as well as novel process concepts in a short time. A major difficulty in the assessment of novel pathways is the connection of different reaction steps to a consistent process pathway and thereby identify missing process steps as well as viable options, e.g., for a specific separation task.

These separations need to be addressed systematically as the choice, type and efficiency of a separation strongly correlates with the mixture composition, the separation task and, thus, the feed composition specification of a subsequent reaction.

2.8 Requirements for a novel screening method

Hence, a systematic identification of required separation steps is mandatory. For an accurate and fast evaluation of the separation effort, the applied models should rely on rigorous thermodynamics. Furthermore, the models need to be independent of design decisions. Design decisions are for example the number of column stages or the feed stage position. The determination of the optimal design parameter either requires a high number of simulation studies or the solution of complex and highly nonlinear optimization problems. Therefore, separation models which are independent of such design decisions are easier to solve. To keep the effort of data collection and property estimation low, the need for additional property data should be as low as possible. Since the RNFA forms a valuable basis for the assessment of novel and existing pathways, the RNFA is extended in this thesis to the integration of separations.

To benefit from the additional knowledge gain by the inclusion of separations, the economic and sustainability criteria need to be revised. The novel criteria should be easy to apply, meaningful in order to achieve the desired target of increasing the sustainability of chemical processes and the criteria should consider the additional knowledge gain due to the inclusion of separations. Furthermore, the criteria should be widely applied and well-known to enable an easy and fair comparison with literature case studies. This allows a validation of the results for known pathways and a statement on the reliability of the method.

Finally, the method should be capable of analyzing the full value chain from biomass transportation via process pathway screening to product-portfolio selection in order to identify viable process concepts. In the following chapter, the method is fully described.

Chapter 3

Process Network Flux Analysis

The gap between the existing mass balance-based evaluation of reaction pathways by means of the RNFA on the one hand and detailed conceptual process design using rigorous methods, on the other hand, is bridged by PNFA. The PNFA can address and evaluate reaction conditions, feasibility and efficiency of separations as well as the process energy requirement of pathways with respect to their economic efficiency and sustainability. While established metrics of the RNFA can be transferred, additional metrics like the energy requirement or global warming potential of pathways are only available, when utilizing the PNFA. Hence, compared to the RNFA, a more detailed analysis of pathways at early stage is enabled.

In the following, the extension of the RNFA to the PNFA methodology is explained. Subsequently, the assessment of the process energy demand as a function of reaction conditions, their influence on the process performance as well as the feasibility and effort of separations are presented. The possibility of an internal energy supply by residue combustion and heat integration are outlined. Afterwards, economic and green chemistry metrics and their applicability in the PNFA are briefly discussed. Furthermore, a model extension of the PNFA covering the supply chain and market modelling are demonstrated. Finally, bi-objective optimization problems are formulated.

3.1 PNFA development

The basis for PNFA is again a reaction network covering all the possible reactions from the reagent(s) to the various products. The extension of the RNFA to PNFA includes the effect of solvents, concentration and separations, which translate the flux level into the respective process level.

3 Process Network Flux Analysis

The following generic reaction r, which converts the reactants C_1 and C_2 into the products C_3 and C_4, is used to demonstrate the differences between RNFA and PNFA:

$$C_1 + C_2 \longrightarrow C_3 + C_4.$$

Figure 3.1 visualizes the flux (left-hand side) and the corresponding process level (right-hand side), for both the existing RNFA methodology and the extension to the PNFA. Similar to the RNFA concept Voll and Marquardt (2012a), the flux level is visualized by nodes and arcs. Nodes represent components and arcs the reactions (RNFA) or processing steps (PNFA), respectively. A reaction can either represent a single conversion or a series of multiple reactions, which take place in the same vessel at the same operating conditions. The literature data on the reactions is used here as a decisive criterion.

For the same bio-chemical conversion, a multitude of data-sets might exist varying in the reaction yield, catalyst, operating conditions, and solvents. Since not only the reaction yield determines the process efficiency, but also the type and efficiency of the separations, multiple data-sets can be represented as multiple parallel nodes.

In the basic RNFA configuration (Fig. 3.1 a) the incoming reactants are supplied by previous reaction fluxes $f_{r,01}$ and $f_{r,02}$. While the product vector **b** represents the network sink, the reactants and products of a reaction might as well further react in any parallel or successive reaction $\mathbf{f}_x$. Based on the general RNFA assumptions (conversion limitation, ideal separations), the translation of the reaction fluxes into specified unit operations at the process level (Fig. 3.1 (a), right-hand side) results in reaction R only, since mixing and separation is assumed to be ideal and to happen spontaneously. Therefore, the molar reaction flow $\dot{n}_r$ is defined as the molar flow of the reaction products. Unconverted reactants and products, which do not react in any subsequent or parallel reaction, are balanced as molar flows $\dot{n}_b$.

Figure 3.1 b) presents the extension of the methodology and therefore the basic configuration of the PNFA. The extension covers the introduction of mixtures and separations allowing a quantification of the separation effort. Fluxes thereof are represented by arcs as well. The mixtures are implemented as pseudo-components and visualized by C_1C_2 and C_3C_4 for the reactant and product mixture, respectively. As for most reactions the literature reports yields close to the theoretical maximum, unconverted reactants play only a minor role in the determination of the separation effort hence an ideal separation of the mixture C_1C_2 from the product mixture C_3C_4 is assumed. The unconverted reactant C_1C_2 might either be recycled to increase the overall conversion (dashed line) or used in any subsequent or parallel reaction.

3.1 PNFA development

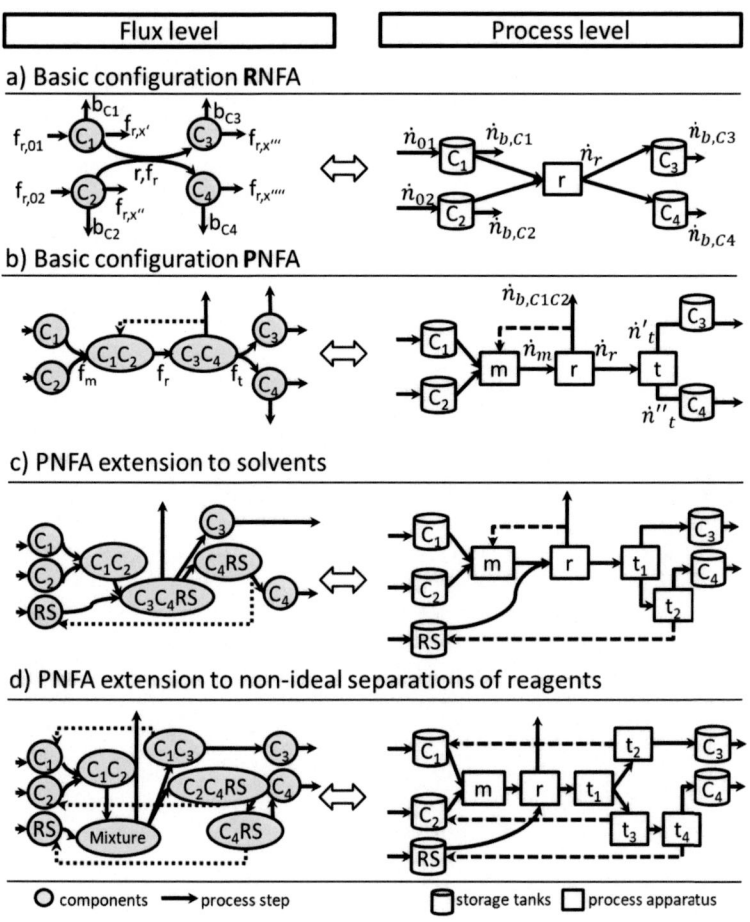

Figure 3.1: The flux level (left-hand side) and the process level (right-hand side) for (a) the basic configuration of the RNFA and (b) the PNFA are presented as well as (c) the extension of the PNFA to include reaction solvents and (d) the extension considering a non-ideal separation of the reagents for an exemplary separation sequence. For simplicity reasons, no energy streams are included in the schemes.

3 Process Network Flux Analysis

If a product mixture consists of more than two products, a systematic determination of the separation scheme exhibiting the minimum energy demand is required. Consequently, separation variants are generated and evaluated using thermodynamically-sound separation models, analogous to the approach of Kraemer et al. (2009) in their synthesis framework. The process energy demand of the particular unit operations is then calculated as constraints based on the fluxes for mixing $\mathbf{f}_m$, reaction $\mathbf{f}_r$ and separation $\mathbf{f}_t$, which imply the amount as well as the composition of the corresponding molar flows. Considering non-idealities, the product mixture C_3C_4 is separated according to pure component and mixture properties. The configuration of the flux level is transferred to the process level, which now consists of a mixer, reactor and separator.

Figure 3.1 c) shows the PNFA configuration for reactions in a solvent. The pseudo-component of the reactants are premixed, converted and mixed with the reaction solvent (RS) to the product mixture C_3C_4RS assuming again an ideal separation or full conversion of the reactants. In order to avoid losses of the solvent during reaction, a reaction yield of unity is assumed for the solvent. In the setup of the matrix, the solvent is added to the product mixture to account for an increased complexity and effort of the separation scheme. While this is important for the notation of the stoichiometric matrix, the translation into the process level (cf. Figure 3.1 c) right) shows a known scheme of premixing the reactants and solvent, followed by a reaction and separation. As the introduction of solvents enhances the purification complexity, a systematic analysis of separation variants is mandatory, if more than one product is diluted in the solvent. After purification the solvent is recycled to reduce the external supply of the respective solvent. The recycles are visualized as dashed lines.

While often high yields are achieved in the context of lignocellulosic biorefineries, the opposite case is possible, e.g., in case of equilibrium-limited reactions. A recycling of reactants is possible, but the influence on the purification strategy needs to be considered. The complexity of the required separation scheme and the number of different sequencing possibilities increases. Further non-idealities like azeotropes can render a separation infeasible or at least inefficient. The general scheme remains the same although additional effort is required to analyze the separation sequences, which is shown for a non-ideal separation in Figure 3.1 d).

Furthermore, the PNFA is based on the following general assumptions. For gaseous reactants, no mass transfer limitations are considered. Gaseous components (e.g., hydrogen, carbon dioxide or carbon monoxide), which are formed during a reaction either have a low solubility in the reaction solvent and are therefore already present in the gas phase or can be easily removed by a pressure relief after the reaction.

3.1.1 Flux balance

Compared to RNFA, the model formulation of PNFA is based on additional fluxes for mixing and separation. In principle, multiple fluxes for mixing and separation might occur per reaction step, e.g., in case a reaction solvent is required (Figure 3.1 c-d). Consequently, the flux balance of the RNFA (Equation 2.1) needs to be adapted for the PNFA as shown in Equation 3.1:

$$\boldsymbol{A} \cdot \boldsymbol{f} = \boldsymbol{b}, \quad \boldsymbol{A} = \begin{bmatrix} \boldsymbol{A}_1 & \boldsymbol{A}_2 & \boldsymbol{A}_3 & \boldsymbol{A}_4 \end{bmatrix}, \boldsymbol{f} = \begin{bmatrix} \boldsymbol{f}_1 \\ \boldsymbol{f}_2 \\ \boldsymbol{f}_3 \\ \boldsymbol{f}_4 \end{bmatrix}. \quad (3.1)$$

The general form of the flux balance remains the same, but the stoichiometric matrix **A** and the flux vector **f** are extended covering now the sub-matrices and the respective fluxes for mixing ($\boldsymbol{A}_3$, $\boldsymbol{f}_3$) and separation ($\boldsymbol{A}_4$, $\boldsymbol{f}_4$).

In order to quantify the separation effort, pseudo-components need to be introduced which represent the mixtures occurring in a process. These pseudo-components are included in the stoichiometric matrix **A** as additional components. The composition of these pseudo-components equals the relationship of the stoichiometric coefficients ν for the specific reaction. In case of the generic reaction the pseudo-component $\nu_{C_1}C_1\nu_{C_2}C_2$ is simply C_1C_2 as all ν are equal to one. If solvents are required, the composition is determined based on the concentration, which is converted into a molar composition and implemented as stoichiometric coefficient in matrix **A** as well. The properties like the molar mass of these pseudo-components are derived based on pure component properties and the molar composition of the pseudo-component, respectively.

Analogous to the RNFA, the process fluxes are constrained by yields (Equation 2.2), while in conceptual design reaction kinetics are often used. However, these are rarely available for novel reactions. The use of reaction kinetics leads to high non-linearities especially if multiple reactions are connected. Thus, only local solutions can be found. The conceptual design method also requires an elaborate initialization procedure as the model is difficult to solve. Therefore, the solvability is simplified using yield parameters instead. However, the main reason to work with yield parameters is the absence of reaction kinetics. While experimental yields are taken as input parameter for the reactions, yields for mixing are set to unity as no losses occur and in case of sharp splits, yields for separation fluxes are set to unity as well, but are easily adaptable in case non-sharp splits occur.

These extensions of RNFA to PNFA result in a formulation that can be interpreted as a superstructure and more specifically as a state task network design (Kondili et al., 1993; Yeomans and Grossmann, 1999). Herein, the tasks are represented by the fluxes for mixing, reaction and separation which are connected via the stoichiometric matrix. The states are represented by the components. Thus, the mathematical basis is similar to existing superstructure approaches (cf. Section 2.3). In contrast to the existing superstructure approaches, the novel PNFA approach allows a fast screening and evaluation of novel process concepts independent of design decisions.

3.1.2 Input data

Both screening approaches rely on a similar network formulation. Thus, the same reaction (stoichiometry, yield) and property data (e.g., molar mass) utilized in the RNFA are required as input for the PNFA. Section 2.4.1 describes a systematic way of retrieving the reaction data and estimating missing pure-component properties.

The PNFA needs additional data like the reaction conditions, i.e., temperature, pressure, concentrations and solvent. For a first feasibility check of the separations, additional pure-component properties are mandatory. Examples are the density, boiling and melting point temperatures or the octanol-water partition coefficient. If the data is not available in literature, missing properties can be estimated using group-contribution methods (e.g., Joback and Reid, 1987; Gani et al., 1997; Marrero and Gani, 2001). Furthermore, mixture properties are required, e.g., to determine vapor-liquid equilibria. Thus, the NRTL model as activity coefficient model is applied. In Section 2.5 a discussion on the choice of the activity coefficient model is presented.

In this thesis, the activity coefficients are retrieved from the Aspen database version 7.3 or are estimated using COSMO-RS for novel molecules and mixtures thereof (Klamt, 2005). Isobaric temperature-dependent activity coefficients are obtained using COSMO-RS and are then utilized to determine the NRTL parameter. For this purpose, the error between the activity coefficients obtained with COSMO-RS and those derived using the NRTL parameter, is minimized. COSMO-RS also provides information about azeotropes and miscibility gaps, which is used to verify the estimated NRTL parameter. In principle, phase stability has to be evaluated. For this purpose, Bollas et al. (2009); Mitsos et al. (2009) and Glass et al. (2017) developed a bi-level approach to obtain a global solution of the parameter estimation problem and depict phase stability correctly. However, since the COSMO-RS data are purely predictive and do not rely on experimental data, the COSMO-RS results itself show uncertainties. Therefore, the bi-level approach is not utilized in this thesis.

The uncertainties arising from using COSMO-RS as basis instead of experimentally validated NRTL parameters is difficult to evaluate, as a consistent thermodynamic model is required to run the separation models (cf. Section 2.7). In order to avoid thermodynamic inconsistencies, an in-depth uncertainty analysis introduced by using COSMO-RS is not conducted. Instead, the specific energy demand for each separation is varied in a sensitivity analysis to address both the uncertainties in the underlying thermodynamic as well as in the separation model. Hence, the utilization of COSMO-RS is helpful to generate NRTL parameters, but the resulting uncertainties cannot be fully determined without extensive experimental studies of the analyzed mixtures. The problem is similar for group-contribution approaches.

Overall, the additional property data required in the PNFA is either available in literature or can be estimated. The information on the reactions is available anyway. Thus, the amount of additional data required for the PNFA, is kept to a minimum.

3.2 Evaluation of processing pathways

The different pathways can be evaluated based on various methods and metrics. Besides a classical mass-based evaluation of processes, the available metrics are classified into the i) thermodynamic, ii) economic and iii) sustainability analysis of networks.

A thermodynamic analysis and feasibility check by minimization of the Gibbs Free Energy is often used in the context of metabolic flux analysis (e.g., Mavrovouniotis, 1991; Jankowski et al., 2008; Henry et al., 2007, 2010; Yim et al., 2011; Woolston et al., 2013), while also first attempts of evaluating biorefinery pathways based on an exergy analysis have been proposed (Frenzel et al., 2014), but the focus of this work lies on the tradeoff between the economic efficiency and sustainability of pathways.

As the energy requirement strongly influences the process performance, the assessment of the process' energy demand is the basis for an efficient and meaningful screening of processing pathways (cf. Section 3.2.1). Furthermore, the energy demand is required to determine the economic efficiency and sustainability. The economic viability of a process concept is addressed as early as possible to prevent non-realistic process concepts, which are not realizable in short or even in long term (decades). The development of sustainable processes, which is especially important in the context of biorefineries, might lead to different process concepts and pathways. Hence, evaluation criteria for both - the economic efficiency (Section 3.2.2) as well as the sustainability of pathways are presented (Section 3.2.3). Finally, an overview of existing allocation methods and their application in the PNFA is given in Section 3.2.4.

3.2.1 Energy demand of pathways

In the following, an approach is presented to determine the energy requirement of a process. Ideal mixing is assumed, such that the energy contribution from the enthalpy of solution is neglected. Therefore, the energy demand of a process consists of the energy duties for reaction and separation. The energy demand can be reduced by an internal energy supply resulting from residue combustion or by means of heat integration.

3.2.1.1 Reactions

The energy requirement for reactions is dominated by the pressure, temperature and phase change prior to a reaction and the enthalpy of reaction during the reaction itself. Previous attempts considered the enthalpy of reaction Δh_R only for a ranking of the pathways (Andiappan et al., 2015), but this contribution also covers the enthalpy of vaporization Δh_{LV} in case of phase changes. These phase changes occur, whenever there is a change from a liquid phase reaction or separation to a gas phase reaction ($\Delta h_{LV} \geq 0$) or vice versa ($\Delta h_{LV} < 0$). Similarly, the phase change from solid to liquid (Δh_{SL}) is introduced. The influence of a temperature change is not considered, as the resulting energy demand is generally low compared to the energy demand for phase change. The reason for this is that only minor temperature changes occur in the context of biorefineries. This way the number of required input property parameters is reduced. This leads to the overall heat requirement $E_{heat,r}$ of the reactions

$$E_{heat,r} = \sum_{j=1}^{N_r} f_j (\Delta h_{LV,j} + \Delta h_{SL,j} + \Delta h_{r,j}) \qquad \Delta h_{LV,j},\ \Delta h_{SL,j},\ \Delta h_{r,j}\ all\ \geq\ 0. \quad (3.2)$$

In case of exothermic reactions ($\Delta h_{r,j} < 0$) or a required condensation prior to a reaction, the energy duties are assigned to the cooling requirement of the reactions

$$E_{cool,r} = -\sum_{j=1}^{N_r} f_j (\Delta h_{LV,j} + \Delta h_{SL,j} + \Delta h_{r,j}) \qquad \Delta h_{LV,j},\ \Delta h_{SL,j},\ \Delta h_{r,j}\ all\ <\ 0. \quad (3.3)$$

Both the enthalpies of reactions as well as the enthalpy of vaporization can be determined prior to the network optimization and are therefore entered as parameter into the optimization problem allowing to distinguish between cooling and heat requirements. The electricity demand of reactions consists of the energy demand for a pressure change. Often, pressure levels are not explicitly stated in publications and might be altered during scale-up anyway.

3.2 Evaluation of processing pathways

For liquid phase reactions, which take place at pressures below 100 bar, the energy demand for a pressure change is small compared to the energy demand for separations. Thus, the contribution to the energy demand is neglected for liquid phase reactions. In contrast, the electricity demand for the compression of gaseous components is substantially higher. Therefore, the following equation considers the gas compression for the electricity requirement $E_{elec,r}$ of reactions

$$E_{elec,r} = \sum_{j=1}^{N_r} f_j \cdot \frac{1}{\eta_{Comp}} \frac{\gamma}{\gamma - 1} RT((\frac{p_j}{p})^{\frac{\gamma-1}{\gamma}} - 1), \tag{3.4}$$

which is based on the compressors efficiency η_{Comp}, a polytropic exponent γ, the ideal gas constant R, the temperature T, the pressure p_j of reaction j as well as the starting pressure p (Biegler et al., 1997).

3.2.1.2 Separations

Mandatory separations are identified, whenever there is (i) a change of the reaction solvent, (ii) an increase in the concentration or (iii) the end of a pathway for the final purification. Gaseous components, like hydrogen or carbon dioxide are directly separated from the liquid mixtures. The identification of suitable and efficient separation strategies for the liquid mixtures influences the overall process performance significantly (Sholl and Lively, 2016). Therefore, a systematic approach is presented which identifies suitable separation types in a first step. In a second step, a fast and reliable assessment of the separation effort is proposed.

Selection of separation techniques A systematic and fast selection of suitable separation strategies is targeted to avoid user-specific selections of separation techniques as well as tedious simulation studies of all alternatives. Therefore, easy-to-apply feasibility criteria are required. Herein, the approach of thermodynamic insights as proposed by Jaksland et al. (1995) is utilized. The idea is based on the driving force concept. This states that separations which take advantage of large property differences of molecules are simple, whereas similar molecules are difficult to separate.

The approach of thermodynamic insights is based on pure-component property ratios between binary pairs which represent the driving force of a separation. For a property ratio equal or close to unity, the driving force vanishes, which renders the separation challenging or even impossible. Multi-component mixtures are subdivided into binary mixtures for the analysis (Jaksland et al., 1995). This method enables a systematic and fast screening of various separation techniques.

Furthermore, only pure component properties are required for the analysis, which are easily accessible via literature search or group-contribution methods and software tools (Joback and Reid, 1987; Gani et al., 1997; Marrero and Gani, 2001).

The analysis of separation techniques is the first step and, thus, conducted prior to the network optimization. The second step is then the determination of the separation effort for all feasible separation techniques. The PNFA considers evaporation, (azeotropic) distillation and extraction as potential separations. For flash evaporation and distillation, the property ratios of the boiling point temperature T_B and vapor pressure p_{oi} are utilized as criteria. For extraction, the octanol-water partition coefficient $\log K_{OW}$, the molar volume V_M and Hildebrandt solubility HS are applied. Lower bounds on the respective property ratios are given in literature (Jaksland et al., 1995; Holtbruegge et al., 2014). If these bounds are exceeded, a separation is feasible. A special case is the analysis of the $\log K_{OW}$ for mixtures with water. Since the $\log K_{OW}$ is not defined for mixtures with water, pure-component $\log K_{OW}$ are given instead and marked in parenthesis in this work. In these cases, the lower bound is set to zero per definition.

For the complex purification of fermentation broths, the list of separation techniques is extended. Precipitation and electrodialysis are utilized for all dissociated fermentation products. Hence, the dissociation degree is a necessary criterion. The dissociation degree is calculated using the logarithmic acid dissociation constant and the pH value of the fermentation. An electrodialysis requires electricity, whereas precipitation results in a stoichiometric amount of gypsum, which needs to be disposed and cannot be further sold due to inevitable contaminations caused by nutrients required for the fermentation (López-Garzón and Straathof, 2014).

Since fermentation products are often highly diluted in water, a reverse osmosis can be applied to increase the concentration and, thus, reduce the effort in a subsequent purification. For these kind of membranes, the molecular weight is applied as feasibility criteria (Jaksland et al., 1995). Furthermore, membrane applicability is restricted from construction to osmotic pressures below 80 bar (Melin and Rautenbach, 2007). Even though this cannot be applied as a priori feasibility criteria, a reverse osmosis is only applied for pressures below this upper limit.

In addition to thermal separation and liquid-liquid extraction, crystallization is considered for the final purification step. For a cooling crystallization, a temperature-dependent solubility of the fermentation product in water is required. If the gradient of the solubility curve is not high enough, crystallization is not a viable option. A lower bound on this gradient is molecule specific and therefore not available. Instead, the enthalpy of fusion is taken as feasibility criteria (Jaksland et al., 1995).

3.2 Evaluation of processing pathways

Separation effort For all feasible techniques, the effort in terms of energy or auxiliary demand is determined prior to the optimization. An overview of existing models for conceptual design of distillation-based separation processes for non-ideal and azeotropic mixtures is given by Skiborowski et al. (2013). Here, the Rectification Body Method (RBM) of Bausa et al. (1998) is used to determine the minimum energy demand of distillation since there is no need for a detailed column specification as input.

For all three and four-component mixtures, various separation sequences are analyzed and the sequence exhibiting the lowest energy demand is used. Azeotropic mixtures are separated either by pressure-swing distillation or by exploiting a miscibility gap in a heteroazeotropic distillation. Bausa et al. (1998) report only minor deviations of the RBM compared to rigorous Aspen models for the application of ideal and complex mixtures. This validation proves the applicability of the RBM for the early-stage screening of the PNFA, whereas uncertainties arising from the model are covered by a sensitivity analysis (cf. Section 2.7).

The minimum solvent demand for extraction is calculated using a thermodynamical sound pinch-based extraction model (Redepenning and Marquardt, 2017). Both separation models only need thermodynamic mixture properties and the feed flow and composition as input, assuming sharp splits. Scheffczyk et al. (2016) even proved the potential and efficiency of these shortcuts for an automated screening of hybrid extraction - distillation processes for more than 4600 solvents. While the models for distillation or extraction are straightforward, simplified models are introduced for the residual alternatives, which are listed in Appendix A.

For all feasible separation techniques, the separation effort in terms of energy or auxiliary demand is determined prior to the optimization. Alternative separation strategies are introduced as separate fluxes, which enables the solver to choose the best performing setup during the screening. Thus, in principle any unit operation can be introduced if the corresponding auxiliary or energy requirements are either known or can be calculated.

The energy duties are parameters in the optimization problem, while the auxiliary demand is covered by the flux balance. This approach does not influence the optimization results as the separation feed composition is fixed but only its mass flow varies depending on the selection of pathways. In addition, it has the advantage that nonlinearities arising from the underlying thermodynamics are decoupled from the optimization problem enabling a simplified solution procedure.

The heat demand of the separations $E_{heat,t}$ is determined based on the sum of the specific energy demands E_{spec} of a separation t,

$$E_{heat,t} = \sum_{j=1}^{N_t} f_{t,j} E_{spec,heat,t}. \tag{3.5}$$

The cooling requirement $E_{cool,t}$ is calculated as

$$E_{cool,t} = -\sum_{j=1}^{N_t} f_{t,j} E_{spec,cool,t}. \tag{3.6}$$

Similarly, the electricity demand of a separation can be described as follows

$$E_{elec,t} = \sum_{j=1}^{N_t} f_{t,j} E_{spec,elec,t}. \tag{3.7}$$

In case of a distillation column, the specific heat demand then reflects the reboiler duty, while the specific cooling demand is given by the condenser duty. In case of heat-integrated columns, the specific electricity demand depicts the work required in a compressor (cf. Section 3.2.1.3). The solvent required for extraction is introduced in the supply matrix $\mathbf{A}_1$. The solvent amount is entered in matrix $\mathbf{A}_3$ in relation to the separation mixture. The composition of the extract and raffinate are given in matrix $\mathbf{A}_4$. The procedure is the same for the residual separation alternatives (precipitation, electrodialysis, reverse osmosis, cooling crystallization). An auxiliary demand is introduced in the matrix $\mathbf{A}$, whereas specific energy duties are covered by the above described equations. A description of the residual separation alternatives is given in Appendix A.

Further separation options are applicable, especially for the purification of fermentation broths. Examples are adsorption and ion exchange, reactive extraction, aqueous two-phase extraction or chromatography (Kurzrock and Weuster-Botz, 2010; López-Garzón and Straathof, 2014; Yenkie et al., 2016). Since modelling these separations requires very specific laboratory data, tailored to the individual components and conditions, the performance thereof are challenging to predict at early stage.

3.2.1.3 Internal energy supply

The external energy demand of processes might be reduced by energy integration. One possibility is the internal energy supply (IES) by the combustion of waste residues.

3.2 Evaluation of processing pathways

An example is the lignin fraction as an efficient lignin exploitation and material usage are still open questions.

The determination of the resulting energy surplus E_{IES} is described by

$$E_{IES} = -\eta_{steam} \cdot \sum_{j=1}^{N_w} f_{w,j} \Delta h_{w,comb}, \qquad (3.8)$$

where additional fluxes f_w are introduced for the combustion of waste residues w and multiplied by the enthalpy of combustion $\Delta h_{w,comb}$ of these residues (Humbird et al., 2012). In a first step, steam is generated which then drives the turbine for electricity production in the second step. Therefore, the energy reduction potential considers losses by the efficiency for steam generation η_{steam}. The generated steam $E_{heat,IES}$

$$E_{heat,IES} = E_{IES} \cdot \zeta \qquad (3.9)$$

and electricity $E_{elec,IES}$

$$E_{elec,IES} = E_{IES} \cdot \eta_{elec} \cdot (1 - \zeta) \qquad (3.10)$$

are then determined by splitting the steam and electricity production using the split fraction ζ. If only steam is generated, the split fraction equals one, while for pure electricity generation, the split fraction equals zero. However, as steam is required for the operation of the subsequent turbine, even in case of electricity generation only, a boiler for steam production as well as the turbine are required. Losses in the turbine, are taken into account by the efficiency for electricity generation η_{elec}. Both the efficiency for steam (η_{heat}=89%) as well as for electricity production (η_{elec}=44%) are derived from primary energy factors for heat e_{heat} (1.12) and electricity e_{elec} (2.28), respectively (International Institute for Sustainability Analysis and Strategy, 2016).

3.2.1.4 Heat integration

An alternative way for energy reduction is heat integration to transfer excess heat from a source (hot stream), which needs to be cooled, to a sink (cold stream), which needs to be heated.

Prior to the development of a heat exchanger network, the heat integration potential and minimum external utility requirement can be determined using the pinch analysis, which decomposes the stream temperatures into temperature intervals TI based on a minimum allowable temperature difference ΔT_{min}. Based on the hot and cold streams present in a temperature interval, the exchanged heat is determined.

The entries of the resulting heat cascade are accumulated. The external heat demand then equals the maximum entry, which needs to be added such that the flow over the pinch temperature T_{pinch} is zero. The cooling demand equals the entry in the lowest temperature interval (Linnhoff, 1993; Biegler et al., 1997).

In the following, the integration of a simultaneous pinch analysis into the process optimization problem is presented with the objective of reducing the external energy demand. To limit the complexity of the resulting problem, the temperature intervals are determined prior to the optimization. For this purpose, the temperature levels of all reactions and separations are collected. In case of hot streams, the specific temperatures are lowered and cold stream temperatures are raised by $\frac{\Delta T_{min}}{2}$. The shifted temperatures are written in a descending order and assigned to temperature intervals. In the next step, a matrix **HI** is set up with the rows referring to the temperature intervals and columns to the reactions and separations. Specific heat or cooling requirements of reactions or separations are entered as parameters partitioned among the applicable temperature intervals. Parameters for the cooling requirements, e.g., a condenser required for a distillation column or an exothermic reaction, are entered with a positive sign, while the specific heat demands, e.g., a column reboiler or an endothermic reaction, are included with a negative sign. These first two steps are carried out prior to the optimization since all required data like the specific heat and cooling requirements as well as the temperature level of every reaction and separation are known parameters. To obtain a fast pinch analysis, adaptations of the operating conditions are not considered at this early stage.

The specific energy requirements in matrix **HI** are multiplied with the reaction and separation fluxes f during optimization leading to the non-accumulated heat cascade **HC**:

$$\mathbf{HI} \cdot \mathbf{f'} = \mathbf{HC}. \qquad (3.11)$$

The entries of **HC** need to be summed up to yield the accumulated heat cascade $\mathbf{HC}_{sum}$. The starting temperature interval is the one exhibiting the highest temperature. Then every entry of $\mathbf{HC}_{sum,TI}$ consists of the previous $\mathbf{HC}_{sum,TI-1}$ entry and the TI specific entry of the heat cascade **HC**, which can be formulated as

$$\mathbf{HC}_{sum,TI} = \mathbf{HC}_{sum,TI-1} + \mathbf{HC}_{TI}. \qquad (3.12)$$

For the final heat cascade $\mathbf{HC}_{final}$ the sum of the accumulated heat cascade $\mathbf{HC}_{sum}$ and the external utility demand $\mathrm{E}_{heat,HI}$

$$\mathbf{HC}_{final,TI} = \mathbf{HC}_{sum,TI} + E_{heat,HI}, \qquad (3.13)$$

is determined. If the heat demand can be fully supplied internally, $\mathrm{E}_{heat,HI}$ equals zero. Otherwise it needs to be specified during optimization.

3.2 Evaluation of processing pathways

In general, $E_{heat,HI}$ refers to the smallest entry of $\mathbf{HC}_{sum}$. Identifying the minimum of vector $\mathbf{HC}_{sum}$ during optimization, leads to a non-smooth and non-continuous function, which is not differentiable and therefore difficult to solve simultaneously during optimization. Alternatively, the following constraint can be added

$$\mathbf{HC}_{sum,TI} \geq 0, \qquad (3.14)$$

which sets a lower bound on $E_{heat,HI}$. The utility requirement is limited to $E_{heat,HI}$. The cooling utility $E_{cool,HI}$ equals the vector $\mathbf{HC}_{final}$ entry of the lowest TI.

The activation of additional pathways, which have no value contribution and are only utilized as heat sources for heat integration, needs to be avoided. An example is an exothermic reaction, which provides heat at a required temperature level sufficient for heat integration, but the subsequent separation is skipped. This way, large amounts of valuable biomass are converted into undesired reaction mixtures. In order to prevent these additional pathways, the entries of the pseudo-components, e.g., for product-solvent mixtures, in the product vector $\mathbf{b}$ are set to zero.

Alternatively to the simultaneous heat integration, the pinch analysis can be conducted subsequently to the process optimization, when all process pathways are fixed. A comparison of the simultaneous and subsequent approach is conducted in Chapter 5.1.6 discussing the additional effort for simultaneous heat integration and the influence on the pathway results compared to the subsequent approach. For the analysis, only the PNFA without residue combustion is considered, since the internal energy supply by combustion leads to an external utility requirement of zero.

The described heat integration potential analysis describes the energy minimization procedure for a fixed ΔT_{min}. However, it is also possible to minimize the TAC prior to heat exchanger design by a variable ΔT_{min}. This approach is called Supertargeting and is useful to determine the influence of the selection of ΔT_{min} on the final cost (Ahmad, 1985; Linnhoff and Ahmad, 1989). Considering in addition the design of heat exchangers and networks thereof, various groups suggested optimization approaches for the design of optimal networks (e.g., Floudas et al., 1986; Yee and Grossmann, 1990; Ciric and Floudas, 1991; Srinivas and El-Halwagi, 1994; El-Halwagi et al., 1995). In this thesis, the heat integration potential analysis is used for an initial assessment of the energy reduction potential. Therefore, Supertargeting and heat exchanger network design are not yet incorporated, but can be easily introduced in the future.

An alternative way for heat integration is to reduce the energy demand of a column by means of vapor recompression (VRC). Here, the distillate is compressed such that the boiling temperature of the distillate is higher compared to the dew point temperature in the reboiler to (partially) supply the energy duty required in the reboiler.

3 Process Network Flux Analysis

Herein, the model of Skiborowski (2014) is applied with a minimum temperature difference of 10 K. As pointed out by Harwardt and Marquardt (2012) VRC is especially useful for close boiling mixtures as the mandatory pressure increase is rather small. Since the investment cost of a compressor correlates with the applied pressure difference (Biegler et al., 1997), these are kept low for close boiling mixtures. Thus, an economic advantage is feasible compared to non-heat integrated setups as major parts of the utility cost are saved.

In the PNFA, both non-heat integrated as well as VRC columns are applied and separate fluxes are introduced enabling the solver to choose the column setup exhibiting either a higher energy demand along with high utility cost but low investment cost or the heat-integrated setup with lower external energy duty but higher investment cost, caused by the additional compressor.

3.2.1.5 Cumulative energy demand

The overall external energy demand depends on the energy requirement for reaction and separations as well as on the heat integration. If heat integration is applied, the binary variable y_{HI} equals one. Hence, the heat requirement E_{heat}

$$E_{heat} = (1 - y_{HI}) \cdot (E_{heat,r} + E_{heat,t}) + y_{HI} \cdot E_{heat,HI} + E_{heat,IES}, \tag{3.15}$$

and cooling duty E_{cool}

$$E_{cool} = (1 - y_{HI}) \cdot (E_{cool,r} + E_{cool,t}) + y_{HI} \cdot E_{cool,HI}, \tag{3.16}$$

consist of the energy demand for all reactions (Eqn. 3.2-3.3), separations (Eqn. 3.5-3.6) and is only altered by means of heat integration (Eqn. 3.11-3.14) and lowered by the contribution from residue combustion (Eq. 3.8). The electricity demand E_{elec} is not influenced by heat integration and, thus, only consists of the requirement for reaction (Eq. 3.4) and separation (Eq. 3.7) reduced by the internal energy supply (Eq. 3.10) according to

$$E_{elec} = E_{elec,r} + E_{elec,t} + E_{elec,IES}. \tag{3.17}$$

The overall cumulative energy demand CED

$$CED = \frac{e_{heat} E_{heat} + e_{elec} E_{elec}}{\sum_{i=1}^{N_{products}} b_{product,i} M_{product,i}}, \tag{3.18}$$

3.2 Evaluation of processing pathways

of a process path is then defined as the sum of all heat (E_{heat}) and electricity requirements (E_{elec}) multiplied with primary energy factors (e) (Curzons et al., 2001). Primary energy factors for a current heat (e_{heat}=1.12) and power plant mix (e_{elec}=2.28) in Germany are utilized (International Institute for Sustainability Analysis and Strategy, 2016). In case the internal energy supply from residue combustion is larger compared to the external heat and electricity duties, the CED turns negative. Therefore, process concepts with a negative CED produce more energy than required and can dispense the surplus to other plants. While a comparison on a mass basis is sufficient for value-added products, a normalization on the heating value and, thus, the enthalpy of combustion $\Delta H_{comb,product}$ is preferred in case of fuels. In these cases CED_{fuel} is therefore applied instead,

$$CED_{fuel} = \frac{e_{heat}E_{heat} + e_{elec}E_{elec}}{\sum_{i=1}^{N_{products}} b_{product,i}\Delta h_{comb,product,i}}. \quad (3.19)$$

For multi-product biorefineries with a co-production of chemicals and fuels, a fair choice of the normalization is difficult. This is further discussed in Section 3.2.4.

3.2.2 Economic efficiency of pathways

In the RNFA only a first estimation of the investment costs based on a correlation to the enthalpy of combustion loss as well as the raw materials have been assessed (Voll and Marquardt, 2012a). With an increased level of detail, the PNFA is capable of addressing the economic efficiency more accurately. The TAC are described by

$$TAC = \sum_{j=1}^{N_s} f_{s,j}M_sP_s + \sum_{i=1}^{N_w} b_{w,i}M_iP_w + \sum_{k=1}^{N_{utility}} E_{utility,k}P_{utility} + \frac{ir}{1-(1+ir)^{-n}}IC, \quad (3.20)$$

which consist of the raw material, waste disposal and utility cost as well as the annualized investment cost. The raw material and waste disposal cost can be determined based on the flux balance (cf. Equation 3.1), the specific prices P_s for the reagents and auxiliaries s as well as for the waste disposal P_w of the residues w and the respective molar mass M. The utility cost are accessed by the respective utility prices $P_{utility}$ for steam, electricity, cooling water and refrigerants and the required energy duties determined by Equations 3.15-3.17. Detailed equipment data is not available at early stage, hence determining the IC is challenging which is further outlined in the following Sections 3.2.2.1 and 3.2.2.2. All price parameters are given in Appendix C.1.

3.2.2.1 Empirical investment costs correlations

The IC of a plant contribute substantially to the overall production cost and reflect the investment risk, leading to the desire of addressing the IC as early as possible in process design. Often in conceptual design, equipment cost estimation methods (e.g., Guthrie's method (Guthrie, 1969)), are utilized, which require detailed design specification like the height and diameter of a distillation column, hence are only applied when the general process concept has been selected. Prior to that stage, early stage screening methods, like the PNFA, are aiming at a relative comparison of different processing pathways and concepts. At this early stage, detailed design specifications for the application of Guthrie's method (Guthrie, 1969) are not known. Therefore, empirical IC correlations are applied (Voll and Marquardt, 2012a; Kokossis et al., 2015; Tsagkari et al., 2016), which are based on less details accepting a higher level of uncertainty (Cheali et al., 2015a).

A number of various empirical correlations have been proposed in literature, which are retrieved from regression to existing processes. As outlined by Tsagkari et al. (2016) none of these methods consider data on biorefineries and therefore a subsequent comparison with existing plants or more detailed process concepts is mandatory to analyze the accuracy of such methods.

The available correlations are summarized in Table 3.1 and are classified into so-called step counting methods as well as heuristic approaches. All step counting methods (Eqn. 3.21-3.32) are based on an empirical pre-factor $Inv1$ and a number of functional units (NFU). A functional unit is herein defined as a significant process step including associated equipment (Petley, 1997; El-Halwagi, 2012; Cheali et al., 2015a). The yearly plant capacity Cap is adopted to the economy of scale rule by the empirical exponent $Inv2$ and a plant conversion factor X specified as mass ratio between desired product and feed input. If not otherwise stated, the correlations are valid for solid and liquid processing, while correlations for gas phase processing (GPP) are marked specifically (Eqn. 3.28-3.29).

Few step-counting methods (Eqn. 3.30-3.32) also consider the process conditions with factors for the temperature F_T, pressure F_p and the material of construction F_{mt}. However, information like the choice of the construction material is often not available at early stage and the nonlinearity is increased as these factors often involve the logarithms of the maximum process pressure or the exponent of the maximum process temperature. While all of these correlations penalize equipment intensive processes and favor large scale plants due to the economy of scale, an improved accuracy of the more detailed models compared to the simple models has not been reported.

3.2 Evaluation of processing pathways

Indeed, Gerrard (2000) analyzed and compared various step-counting methods for an acetic anhydride production and obtained similar IC. Thus, no benefit is obtained, if more details like the operation conditions are taken into account.

An alternative approach proposed by Lange (2001) correlates the IC to the enthalpy of combustion loss $E_{comb,loss}$ (Eq. 3.33) or to the transferred energy within a process $E_{transfer}$ (Eqn. 3.34) (Lange, 2001, 2007). The IC are therefore independent from the plant size or complexity in terms of the number of required equipment. As pointed out by Lange (2001) the deviation from the regression is high for small-scale and heat-neutral reactions, therefore applicability to biorefineries is difficult (Tsagkari et al., 2016). Cheali et al. (2015a) suggest even simpler methods by assuming a general payback period and a contribution of the raw material cost of 80-90% to the total cost (Eq. 3.35), which leads to an IC correlation based on the difference between the plant's revenue TR and the feedstock cost. Alternatively, Cheali et al. (2015a) propose to predict the IC based on the plant's run time (Eq. 3.36). Although these heuristics might apply for conventional fossil-based processes in the production of bulk chemicals, a transfer to novel and innovative process concepts is challenging as biorefinery concepts might be realized under different economic or technical circumstances. (cf. Section 2.1).

Tsagkari et al. (2016) compared the accuracy of the various methods for biorefinery processes with data from existing plants. Most methods predict the IC in the right order of magnitude, though the accuracy varies significantly depending on the case study. Besides determining the IC as accurate as possible at early stage, a relative comparison of equipment intensive pathways with many reaction and separation steps and equipment less pathways like one-pot reactions is the major goal of integrating the IC assessment in the PNFA. The heuristic approaches (Eqn. 3.33- 3.36) are not applicable for a differentiation thereof. Furthermore, the required knowledge for the more detailed step-counting methods (Eqn. 3.30- 3.32) is not available. Therefore, the IC determination in the PNFA is based on a simple step-counting method providing a capacity-dependent and easy to apply correlation. In Chapter 4.1.2 the accuracy of this IC correlation is discussed. In the following, the IC integration into the PNFA and the determination of the NFU is described.

Table 3.1: Overview of empirical investment costs models with IC expressed in US $.

		correlation	Year	Inv1	Inv2	References
step counting methods	(3.21)	$IC^1 = Inv1 \cdot NFU \cdot (\frac{Cap}{X})^{Inv2}$	1978	362 [5]	0.675	Bridgwater and Mumford (1979)
	(3.22)	$IC^1 = Inv1 \cdot NFU \cdot (\frac{Cap}{X})^{Inv2}$	2000	2924 [5]	0.675	Gerrard (2000); Coker and Ludwig (2007)
	(3.23)	$IC^1 = Inv1 \cdot NFU \cdot Cap^{Inv2}$	2010	7000	0.68	El-Halwagi (2012)
	(3.24)	$IC^2 = Inv1 \cdot NFU \cdot (\frac{Cap}{X})^{Inv2}$	1978	31703 [5]	0.30	Bridgwater and Mumford (1979)
	(3.25)	$IC^2 = Inv1 \cdot NFU \cdot (\frac{Cap}{X})^{Inv2}$	2000	256893 [5]	0.30	Gerrard (2000); Coker and Ludwig (2007)
	(3.26)	$IC^2 = Inv1 \cdot NFU \cdot Cap^{Inv2}$	2010	458000	0.30	El-Halwagi (2012)
	(3.27)	$IC = Inv1 \cdot NFU \cdot Cap^{Inv2}$	2000	15999 [5]	0.615	Gerrard (2000)
	(3.28)	$IC_{GPP} = Inv1 \cdot NFU \cdot (\frac{Cap}{X})^{Inv2}$	1978	4885 [5]	0.62	Bridgwater and Mumford (1979)
	(3.29)	$IC_{GPP} = Inv1 \cdot NFU \cdot Cap^{Inv2}$	2010	36000	0.62	El-Halwagi (2012)
	(3.30)	$IC = Inv1 \cdot NFU \cdot Cap^{Inv2} \cdot F_T F_p F_{mt}$	2000	7439 [5]	0.639	Gerrard (2000)
	(3.31)	$IC^3 = Inv1 \cdot NFU \cdot Cap^{Inv2} \cdot 10^{F_T + F_p + F_{mt}}$	2000	11317 [5]	0.60	Zevnik and Buchanan (1963)
	(3.32)	$IC^4 = Inv1 \cdot NFU \cdot Cap^{Inv2} \cdot 10^{F_T + F_p + F_{mt}}$	2000	26180 [5]	0.50	Zevnik and Buchanan (1963)
heuristics	(3.33)	$IC = Inv1 \cdot E_{comb.loss}^{Inv2}$	1993	3000000	0.84	Lange (2001)
	(3.34)	$IC = Inv1 \cdot E_{transfer}^{Inv2}$	1993	2900000	0.55	Lange (2001)
	(3.35)	$IC = 4 \cdot (TR - 1.2 \sum_{j=1}^{N_s} f_{s,j} M_s P_s)$	-	-	-	Cheali et al. (2015a)
	(3.36)	$IC = \frac{2}{5} \cdot n$	-	-	-	Cheali et al. (2015a)

[1] Correlation valid for a capacity $\geq$ 60000 t/y
[2] Correlation valid for a capacity < 60000 t/y
[3] Correlation valid for a capacity $\geq$ 4536 t/y
[4] Correlation valid for a capacity < 4536 t/y
[5] Converted based on an average exchange rate of 1.920 (1978) or 1.515 (2000) US$ per british pound (UK Forex Foreign Exchange, 2016)

3.2.2.2 Assessment of the investment costs in the PNFA

In this work, the most recent step counting methodology of El-Halwagi (2012) is taken for the IC determination of solid and liquid processes (Eq. 3.23) and gas-phase processes (3.29), respectively. The investment cost are updated to the recent year using the Chemical Engineering Plant Cost Index (CEPCI) as outlined in the following

$$IC = \frac{CEPCI}{CEPCI_{2010}} \cdot Inv1 \cdot NFU \cdot Cap^{Inv2}. \tag{3.37}$$

The plant capacity Cap is determined based on the product flux $b_{product}$ and molar mass $M_{product}$

$$Cap = \sum_{i=1}^{N_{products}} b_{product,i} M_{product}. \tag{3.38}$$

The NFU is then specified based on the number of non-zero reactions, separations, heat-integrated separations by means of VRC and a single unit each for the boiler and turbine. In order to apply the big M formulation, an upper limit max is defined such that binary variables y equal one if the respective units are active as shown in the following:

$$f_j \leq f_{max} \cdot y_j, \tag{3.39}$$

$$f_{j,VRC} \leq f_{max} \cdot y_{j,VRC}, \tag{3.40}$$

$$E_{IES} \leq E_{IES,max} \cdot y_{steam}, \tag{3.41}$$

$$E_{elec,IES} \leq E_{IES,max} \cdot y_{elec}. \tag{3.42}$$

In case no internal energy is supplied, neither the boiler nor the turbine are considered and the binary variables y_{steam} and y_{elec} equal zero. Generating steam only, the IC for the turbine are neglected (y_{elec}=0), while for the generation of electricity, both the boiler as well as the turbine contribute to the IC (y_{steam}=y_{elec}=1) as described in Section 3.2.1.3. For the separation fluxes, a non-heat integrated as well as an internally heat-integrated solution by means of VRC is considered (cf. Section 3.2.1.4). For the VRC separation, an IC twice as high is considered in order to account for the increased plant complexity due to the additional compressor. The number of functional units is then determined by summing up the binary variables, leading to the following equation:

$$NFU = \sum_{j=1}^{N_r+N_t} y_j + y_{steam} + y_{elec} + 2 \sum_{j=1}^{N_{t,vrc}} y_{VRC,j}. \tag{3.43}$$

3.2.3 Sustainability of pathways

Sustainable pathways target the elimination or at least reduction of hazardous and harmful components in process design (Anastas and Warner, 1998). While a detailed life-cycle assessment (LCA) covering all aspects from cradle to grave, meaning from the origin of the raw materials to the products' after-life is always envisaged, in most cases the data situation is insufficient. In addition, no standardized procedure for the setup of LCAs exists, as the norm only clarifies the terms and definitions (ISO 14040, 2006). Hence, green metrics have been developed, addressing sustainability based on the green chemistry and green engineering principles (Anastas and Warner, 1998; Anastas and Zimmerman, 2003). Although not all of these metrics are applicable at early stage, their consideration in a subsequent conceptual and detailed process design as well as in the production process and the products' after-life is important. While an elaborate overview of all of these metrics is given by Broeren et al. (2017), herein the system boundaries are set to processes only.

A weighting of single metrics to one environmental impact (Uhlman and Saling, 2010) or an environmental health and safety index (Koller et al., 2000; Arab, 2013; Banimostafa et al., 2015) is often proposed, summing up numerous sustainability criteria to one single factor. However, the required weighting factors are either subjectively chosen or are case study specific. Also, a normalization based on the worst performing pathway is utilized (Uhlman and Saling, 2010), which renders a comparison of different case-studies and of different weighting approaches infeasible (Banimostafa et al., 2012; Arab, 2013; Rajagopalan et al., 2017).

In the following, the applicability of various well-known sustainability metrics for the RNFA, PNFA and conceptual design level are discussed. The RNFA constitutes the first level with the lowest data requirements. The PNFA is then the second and the conceptual design the third level. Since the level of detail is increased stepwise, applicability of criteria of the previous level is always maintained. Table 3.2 summarizes various metrics and sustainability criteria. Non-applicable criteria are marked with "-". Those giving a first rough estimation are marked with a checkmark in parenthesis "($\checkmark$)". If an accurate determination of a criteria is possible, these are marked with a checkmark without parenthesis "$\checkmark$".

The criteria in Table 3.2 are ordered according to the aforementioned green chemistry and engineering principles. According to these principles, the table summarizes criteria, which target the prevention of waste, a high mass efficiency, the utilization of non-hazardous materials, a minimization of auxiliaries and energy and the maximization of renewables (Anastas and Warner, 1998; Anastas and Zimmerman, 2003).

3.2 Evaluation of processing pathways

While criteria of these seven principles are accessible at early stage, aspects of the products after-life, analytical methodologies or safety issues during production (Anastas and Warner, 1998; Anastas and Zimmerman, 2003) as well as criteria like land-use change or risk (Uhlman and Saling, 2010) can hardly be determined at early stage as the mandatory data is not available. These principles are therefore not further discussed, whereas a brief description of the seven applicable principles is given now.

In the category targeting **waste minimization**, most criteria are accessible at the RNFA level. However, as reaction solvents and separations are only considered later in the PNFA, a large discrepancy of the results between both levels is expected. The waste treatment energy can neither be determined at the RNFA nor at the PNFA stage as the calculation requires detailed knowledge on the type and composition of the disposed waste. The waste energy ratio and the global warming potential cannot be analyzed in the RNFA, but with the PNFA as only then the process energy demand along with the attributed global warming potential are determined.

The criteria in the category of **mass efficiency maximization** target a high atom efficiency or high mass yield, hence most of the criteria are applicable at the RNFA level. One exception is the turnover number of a catalyst (TON). Both the RNFA and PNFA address processes based on laboratory data and the TON is often not known at that development stage.

Even more difficult to determine are the criteria associated with the **minimization of hazardous products** ($b_{hazardous}$). A first insight is given by the toxicity score of materials (Uhlman and Saling, 2010), which is correlated to the risk and safety statements of a material and therefore easily applicable. The toxicity of non-classified materials can be estimated based on a molecule with high similarity in a first assessment. However, accuracy of this approach is limited as it is based on six different classifications only, ranging from less severe risks to carcinogenic effects. While the application of the effective mass yield requires knowledge on the hazards of the reagents, which can be estimated based on the aforementioned toxicity score, the metrics on the pollution and ecotoxicity require expertise on the bio-accumulation of the materials and are therefore not applicable at early stage.

3 Process Network Flux Analysis

Table 3.2: Sustainability criteria for the early design stage.

	Criteria	Symbol	RNFA	PNFA	#Eqn.	References
min. waste	E-factor[1]	E-factor	(✓)	✓	B.1	Sheldon, Constable et al. (2002)
	Waste intensity	WI	(✓)	✓	B.2	Jimenez-Gonzalez et al. (2011)
	Waste percentage[2]	WP	(✓)	✓	B.3	McElroy et al. (2015)
	Waste treatment energy	WT	-	-	-	Jimenez-Gonzalez et al. (2011)
	Waste energy ratio	WER	-	✓	B.4	Jimenez-Gonzalez et al. (2011)
	Emissions	Em	✓	✓	B.5	Saling et al. (2002); Uhlman and Saling (2010)
	Global warming potential	GWP	-	✓	3.45	Curzons et al. (2001)
max. mass efficiency	Atom economy	AE	✓	✓	B.6	Trost (1991); Constable et al. (2002)
	Atom utilization	AU	✓	✓	B.7	Ribeiro and Machado (2013)
	Element efficiency	EE	✓	✓	B.8	Ribeiro and Machado (2013)
	Carbon efficiency	CE	✓	✓	B.9	Curzons et al. (2001); Constable et al. (2002)
	Product yield	Y	✓	✓	2.2	Voll and Marquardt (2012a)
	Reaction mass efficiency	RME	✓	✓	B.10	Curzons et al. (2001); Constable et al. (2002)
	Resource consumption[3]	RC	✓	✓	B.11	Saling et al. (2002); Uhlman and Saling (2010)
	Generalized RME[4]	gRME	-	✓	B.12	Augé (2008)
	Mass intensity[5]	MI	-	✓	B.13	Curzons et al. (2001); Constable et al. (2002)
	Process mass intensity[2]	PMI	(✓)	✓	B.14	Jimenez-Gonzalez et al. (2011)
	Optimum efficiency[2]	OE	✓	-	B.15	McElroy et al. (2015)
	Turnover number catalyst	TON	-	-	-	Bourdart (1995)
	Space time yield	STY	-	✓	B.16	McElroy et al. (2015)

50

3.2 Evaluation of processing pathways

	Criteria	Symbol	RNFA	PNFA	#Eqn.	References
min. hazardous potential	Toxicity Potential	ToxP	✓	✓	B.17	Saling et al. (2002); Uhlman and Saling (2010)
	Effective mass yield	EMS	(✓)	(✓)	B.18	Hudlicky et al. (1999)
	Pollution/Bioaccumulation	PB	-	-	-	Curzons et al. (2001)
	Ecotoxicity	EcoTox	-	-	-	Curzons et al. (2001)
min. auxiliaries	Hydrogen requirement	H2	✓	✓	B.19	Voll and Marquardt (2012a)
	Solvent intensity	SI	-	✓	B.20	Curzons et al. (2001)
	Solvent recovery (energy)	SR	-	✓	B.21	Curzons et al. (2001), Gonzales2011
	Solvent energy ratio	SER	-	✓	B.22	Jimenez-Gonzalez et al. (2011)
min. energy	Energy efficiency combustion	EC	✓	✓	B.23	Voll and Marquardt (2012a)
	Energy efficiency formation	Eform	✓	✓	B.24	Voll and Marquardt (2012a)
	Cumulative energy demand	CED	-	✓	3.18	Curzons et al. (2001)
	Life cycle energy	LCE	-	-	-	Jimenez-Gonzalez et al. (2011)
max. renewables	Renewables intensity	RI	✓	✓	B.25	McElroy et al. (2015)
	Renewables percentage[2]	RP	(✓)	✓	B.26	McElroy et al. (2015)
-	Number of reaction steps	#	✓	✓	-	Voll and Marquardt (2012a)
	Number of processing steps	#	(✓)	✓	-	

[1] Alternative names are environmental quotient (Andraos, 2005) and mass loss index (Heinzle et al., 1998).
[2] Redundant criteria, which can be calculated based on a combination of non-redundant criteria.
[3] Alternative name is mass index (Eissen and Metzger, 2002).
[4] Alternative name is process mass efficiency (Jimenez-Gonzalez et al., 2011).
[5] Reciprocal is called mass productivity (Jimenez-Gonzalez et al., 2011).

3 Process Network Flux Analysis

In contrast, **minimization of auxiliaries** ($f_{auxiliaries}$) can be targeted already at the RNFA level in case a stoichiometric requirement of a co-reagent, like hydrogen, is necessary. In order to determine the solvent amount for either reaction or separation (e.g., extraction, absorption), these steps need to be specified explicitly. Thus, the solvent intensity, solvent recovery energy and solvent to energy ratio are accessible at the PNFA level.

For the same reason, the process energy requirement in the category **energy minimization** cannot be calculated at the RNFA level. Thus, the RNFA covers only the energy efficiency of combustion and formation, which are derived based on the stoichiometry of a reaction. The life-cycle energy demand requires even more detailed knowledge on the reagents supply and the products' after-life and cannot be applied neither in the RNFA nor in the PNFA.

Finally, the **maximization of renewable** (f_{bio}) and number of reactions and processing steps are easily accessible at early stage. All equations for applicable metrics are referenced and summarized in Appendix B.

In this thesis, mainly the performance of transportation fuels as well as value-added co-products are analyzed. In order to allow a direct assessment of the methodological advancement from RNFA to PNFA, an adapted formulation of the EI is utilized as RNFA sustainability criteria in this work. The adapted environmental impact EI' allows a fair comparison of various case studies and circumvent an iterative calculation procedure (cf. Section 2.4.3). In the formulation for the adapted EI',

$$EI' = wc_{Ec}Ec + wc_{RC}RC + wc_{Em}Em + wc_{ToxP}ToxP, \qquad (3.44)$$

the weighting criteria wc are calibrated such that in case of an annual ethanol production of 100,000 t all factors contribute equally to the EI'. Ethanol is taken as a reference with an EI' set to the value of one, because it is an established bio-fuel with a well-known production process (Ulonska et al., 2016b). Fixing these weighting criteria enables an easy and fair comparison of different case studies.

One of the major advantages of the PNFA is that more process information is taken into account which allows a more detailed analysis compared to the RNFA. In the context of biofuels, the energy demand is of special interest as the external utility requirement needs to be lower compared to the product's heating value to ensure a sustainable production (cf. Section 3.2.1.5). Therefore, CED_{fuel} is used in this thesis to address the sustainability of a process.

If additional aspects, e.g., from biomass transportation are considered simultaneously or a literature comparison of the results is targeted, it is translated into the non-renewable global warming potential (GWP$_{process}$)

$$GWP_{process} = \frac{gwp_{heat}(E_{heat} + E_{heat,IES}) + gwp_{elec}(E_{elec} + E_{elec,IES})}{\sum_{i=1}^{N_{products}} b_{product,i} \Delta h_{comb,product}}. \tag{3.45}$$

It is also based on the process energy demand and specific GWP factors gwp for heat (72 $\frac{\text{g CO}_{2,\text{equivalents}}}{\text{MJ}_{\text{heat}}}$) and electricity (171 $\frac{\text{g CO}_{2,\text{equivalents}}}{\text{MJ}_{\text{elec}}}$) in Germany (International Institute for Sustainability Analysis and Strategy, 2016). Therefore, GWP$_{process}$ is used in Section 4.1.3 for a literature comparison and in Sections 5.1.7.2 and 5.3 when transportation and pathway selection are considered simultaneously.

Although the analysis of the non-renewable GWP is chosen, this assumption can be adapted, if, for instance, more than one renewable energy source is available. In addition, the pathways can be evaluated by means of the criteria presented in Table 3.2. In this work, especially the carbon efficiency (Eq. B.9) and the conversion efficiency of the enthalpy of combustion are discussed. The conversion efficiency of the enthalpy of combustion η_{Comb} is defined as

$$\eta_{Comb} = \frac{f_{S1} \Delta h_{comb,biomass}}{\sum_{products}(b_{product} \Delta h_{comb,product})}. \tag{3.46}$$

In case no auxiliaries (e.g., hydrogen) are integrated into the product molecule, η_{Comb} exhibits a lower boundary value of unity, which equals a full conversion of the incoming biomass heating value into the desired products.

3.2.4 Allocation

For multi-product biorefineries, e.g., co-producing fuels and chemicals, the choice of a meaningful functional unit and, thus, the question of splitting the environmental burden between different products arises (ISO 14040, 2006). In principle, it is possible to allocate the environmental burden completely to the main product and neglect the influence of co-product manufacturing (Sandin et al., 2015). This represents, however, a simplification for most cases, as additional effort is required in the purification of the co-products, which also contributes to the overall environmental burden. If an additional process pathway is selected, the influence of co-products to the environmental burden cannot be neglected. Answering the question of allocating the environmental burden to the various products is crucial as it influences the results strongly (e.g., Kim and Dale, 2003; Luo et al., 2009; Kaufman et al., 2010; Ahlgren et al., 2015).

Kaufman et al. (2010) report values of 35 to 76 $\frac{\text{g CO2}}{\text{MJ}}$ for a corn-based (corn-grain and corn stover) ethanol production along with distillers grain and excess electricity.

The differences in the results originate from allocation choices only. These differences translate into 36-79% of the global warming potential of gasoline (Kaufman et al., 2010) which renders a fair comparison of ethanol and gasoline challenging. A similar observation has been made by Luo et al. (2009), who compared the GWP of gasoline with corn-stover ethanol and an excess production of electricity. While for two allocation approaches a decrease in the GWP for ethanol compared to gasoline is achieved (mass and energy allocation), one allocation method shows an increase in the GWP (economic allocation). This illustrates the difficulties of splitting the environmental burden between different products. This also impedes the comparability of different case studies. Therefore, different allocation methods need to be analyzed to obtain a reliable statement.

In the following, allocation methods are briefly reviewed and their applicability in the PNFA discussed. In principle, two major allocation methodologies are available, namely the substitution and partitioning method as well as hybrid approaches thereof (e.g., Kaufman et al., 2010; Cherubini et al., 2011; Ahlgren et al., 2015).

The substitution method takes into account the avoided conventional production of products. The overall environmental burden is reduced by a credit for the avoided processes and charged to the main product only (Cherubini et al., 2011). For an appropriate application of this method, the co-products or at least their product function needs to be established and detailed knowledge of the conventional processes is required. This is in principle possible in case of bio-fuel production, if a whole cradle-to-grave analysis is conducted, considering the different raw materials, their supply, the processes as well as the emissions from combustion. Here, novel biofuels can be compared to fossil fuels but also to established bio-fuels like ethanol or bio-diesel. For the analysis of value-added co-products this is often not feasible as novel molecules are proposed only in the biorefinery context. These processes along with the products are not yet fully understood or established rendering a comparison challenging. Hence, main drawbacks of the substitution method are hypothetical productions leading to a reduction of actual emissions by considering virtual emissions (Kaufman et al., 2010) and an insufficient data situation (Ahlgren et al., 2015).

The second method is the partitioning approach, which partitions the environmental burden between the different products. The partitioning can be conducted based on the products' mass, energy or exergy content as well as on their economic value (Cherubini et al., 2011; Ahlgren et al., 2015). If mass partitioning is chosen, the environmental burden is split up based on the weight fraction of the products, which is applicable as long as no significant energy production is performed.

In contrast, partitioning based on the energy content of the products, is suitable in case of biofuel production, but lacks consistency in the production of chemicals. Analyzing a biorefinery producing biofuels, chemicals and excess energy in parallel, a partitioning based on the exergy content of the product is often chosen. This approach is mainly limited due to a lack of public understanding of exergy (Cherubini et al., 2011). Finally, a partitioning based on the economic value of the products is conceivable as well (Patel et al., 2012). Compelling reasons against an economic partitioning are missing price information for novel molecules and an exposure of LCA results to market and price fluctuations (Sandin et al., 2015; Cherubini et al., 2011).

In the multi-product biorefinery analyzed in this thesis biofuels and valued-added chemicals are produced in parallel (cf. Section 5.3). Since the required knowledge for the application of the substitution method is missing at the early stage of process design, the partitioning method is chosen in a first step. Therefore, partitioning approaches based on the mass flow of the products, the heating value (Δh_{comb}) and exergy content (E_{ex}) are compared and differences in the results are pointed out. GWP allocation for a component i is based on the overall GWP of the process (Eq. 3.45) and the weight (Eq. 3.47), heating value (Eq. 3.48) and exergy (Eq. 3.49) fraction according to:

$$GWP_{allocation,mass,i} = GWP \cdot \frac{b_i M_i}{\sum_{i=1}^{N_{products}} b_{product,i} M_{product,i}}, \quad (3.47)$$

$$GWP_{allocation,energy,i} = GWP \cdot \frac{b_i \Delta H_{comb,i}}{\sum_{i=1}^{N_{products}} b_{product,i} \Delta h_{comb,i}}, \quad (3.48)$$

$$GWP_{allocation,exergy,i} = GWP \cdot \frac{b_i E_{exergy,i}}{\sum_{i=1}^{N_{products}} b_{product,i} E_{ex,i}}. \quad (3.49)$$

In Appendix C.4, a brief description is given for the determination of the exergy content of a molecule.

3.3 Supply chain design

The methodology also covers the biomass supply chain to analyze the effect of transportation on the process viability. Since the case study herein covers a small region and transportation by ship or railway is associated with high non-distance-related cost (Schwaderer, 2012) and limited accessibility, only truck transportation is considered.

In addition, transportation via truck is favorable for small distances of less than 100 - 150 km (Hamelinck et al., 2005; Kappler, 2008). The method can be easily extended to cover alternative transportation modes in the future and thereby identify differences between distributed and centralized production as described by Lara and Grossmann (2016) and You and Wang (2011).

In a first step, the amount of biomass collected at each origin (index o) as well as the total collected biomass need to be identified. The biomass transportation flux $f_{BT,o,d}$ from one harvest site to all destinations (index d) is limited by the maximal biomass capacity available at the origin $Cap_{biomass,o}$,

$$\sum_{d=1}^{N_d} f_{BT,o,d} M_{biomass} \leq Cap_{biomass,o} \quad \forall\ o. \tag{3.50}$$

The underlying assumption herein is that biomass, i.e., residual and small wood, is a homogeneous good, meaning that it has the same characteristics and quality regardless of the time and location it was delivered from. The utilized data by Dieter et al. (2001) does not include information on seasonal biomass composition. Therefore, seasonality is not yet included in the method. The same molar mass $M_{biomass}$ is utilized for all origins o. In a next step, the total amount of biomass collected, f_{S1}, equals the amount from all origins and delivered to all destinations

$$f_{S1} = \sum_{d=1}^{N_d} \sum_{o=1}^{N_o} f_{BT,o,d}. \tag{3.51}$$

Identifying production sites, the big M formulation is used such that the binary variables $y_{site,d}$ equal one if the site d is chosen

$$\sum_{o=1}^{N_o} f_{BT,o,d} M_{biomass} \leq f_{BT,max,d} \cdot y_{site,d}, \tag{3.52}$$

with $f_{BT,max,d}$ representing the maximal amount, which can be delivered to a plant. Summing up the binary variables for the production sites

$$\sum_{d=1}^{N_d} y_{site,d} \leq \lambda \tag{3.53}$$

the maximal allowable number of plant locations λ specifies an upper boundary of production sites. Based on the transport flux and the distances D between harvest and production site, the transportation cost can be determined.

3.3 Supply chain design

The biomass transportation cost consists of a distance-independent price $P_{BT,fix}$ and a distance-dependent price $P_{BT,var}$ which covers the expenses for fuel and wages (Mahmudi and Flynn, 2006; Schwaderer, 2012). These prices are specific for each country and depend on the biomass type and water content, w_{H2O}. In Appendix C.6, an overview is given with known price parameters for the various regions in the world and different biomass types from literature. In general, residual wood causes higher variable prices compared to wood chips due to a lower packing density. Furthermore, the variable prices strongly depend on the region. In this work, the price parameters of Schwaderer (2012) are utilized, which are specific for residual wood in Germany and consider a water content w_{H2O} of 0.5. Herein, the reason to address biomass composition and water content separately is the data on biomass availability reported by Dieter et al. (2001), which is given in dry tons and neither specifies the biomass nor its water content. Considering forest residues specifically, a water content of 50% agrees well with a typical water content of 44% proposed by Mantau (2012). Finally, the transportation price is updated by a factor for logistics prices in Germany called "Verkehrsrundschau" (VR) index from the reference year 2011 to the recent year. The biomass transport cost (BTC) from one harvest to one production site can then be written as

$$BTC_{o,d} = (P_{BT,fix} + P_{BT,var} \cdot D_{o,d}) \cdot \frac{1}{1 - w_{H2O}} \cdot f_{BT,o,d} M_{biomass} \frac{VR}{VR_{2011}}. \quad (3.54)$$

The yearly total biomass transportation cost (TBTC) are obtained by summing up the BTC for all harvest and production sites

$$TBTC = \sum_{d=1}^{N_d} \sum_{o=1}^{N_o} BTC_{BT,o,d}. \quad (3.55)$$

As biomass transportation contributes to the overall global warming potential, GWP is utilized as second objective function. The GWP for biomass transport (GWP_{BT}) is based on a diesel truck specific factor (GWP_{truck}) which correlates with the amount of dry biomass transported and the total distance

$$GWP_{BT} = GWP_{truck} \cdot \sum_{d=1}^{N_d} \sum_{o=1}^{N_o} f_{BT,o,d} M_{biomass} D_{o,d}. \quad (3.56)$$

Different GWP_{truck} factors are given depending on the truck size, reference year, biomass type, density, and moisture content. Börjesson (1996) estimate 0.073 $\frac{kg\ CO_2}{t\ km}$ for 2015, Kumar et al. (2006) consider 0.095 $\frac{kg\ CO_2}{t\ km}$, whereas Garcia and You (2015a) propose a value of 0.18 $\frac{kg\ CO_2}{t\ km}$, all without further specifying the underlying assumptions, truck size or biomass.

3 Process Network Flux Analysis

Based on the transport of wood chips in Germany, Kappler (2008) determines a GWP_{truck} factor of 0.257 $\frac{kg\ CO_2}{t\ km}$ for a water content of 50 wt%. Since this work aims at a supply chain optimization in Germany as well, the GWP_{truck} factor of Kappler (2008) is applied. The values for the biomass capacities $C_{biomass,o}$ are based on dry biomass, hence the GWP_{truck} factor is divided by a factor of two (w_{H2O} of 0.5), leading to a GWP_{truck} factor of 0.129 $\frac{kg\ CO_2}{dry\ t\ km}$, which is used in the remainder of this thesis. The supply chain submodel is linear.

3.4 Market and price modeling

When envisioning a multi-product biorefinery producing both fuels as well as chemicals, the different product value chains need to be considered already in the early design stage, when economic efficiency and realizability of a biorefinery are analyzed. The choice of chemicals and hence production pathways strongly correlate with the value added potential of chosen chemicals. Considering the impact of a new biorefinery process, a simplified supply and demand model is introduced to account for variations in the prices of biomass and chemicals depending on the market sizes. A simple model with a linear price dependency on the market quantity is proposed. Although the model is accompanied by uncertainties, the degree of nonlinearity as well as a need for a vast amount of market data is kept low (Mansoornejad et al., 2011; Malueg, 1994). Since the output of small-scale biorefineries are orders of magnitudes lower compared to the fuel demand, no additional price - market dependency is considered for fuel production. With this first rough market assessment, an integration into early stage process design is enabled.

The price of chemicals is determined based on a current initial price $P_{initial,chem}$, the fraction of current market volume q_{chem} and a price scenario parameter $P_{chem,scenario}$:

$$P_{chem} = P_{initial,chem} - P_{chem,scenario} \cdot q_{chem}. \tag{3.57}$$

Herein, the same market-price dependency for intermediate and special chemicals is assumed as a clear differentiation thereof and a viable market prediction for bio-based chemicals is difficult at the current state of biorefinery development. For commodities (e.g., fuel), no market-price dependency is considered, as the output of a biorefinery is orders of magnitudes lower than the market, such that the influence is negligible.

Market sizes and prices of chemicals are often given for a specific year. To allow a fair comparison of the chemicals, the market sizes herein are calculated for a reference year using the specific component annual growth rates (CAGR).

3.4 Market and price modeling

The prices are updated applying the producer price index for industrial chemicals. A more detailed description is given in Appendix C.6. The biomass price function is built analogously:

$$P_{biomass} = P_{initial,biomass} + P_{biomass,scenario} \cdot q_{biomass}. \qquad (3.58)$$

The fraction of chemicals produced

$$q_{chem} = \frac{b_{chem}}{ms_{chem}}, \qquad (3.59)$$

and biomass consumed by the biorefinery

$$q_{biomass} = \frac{f_{S1}}{ms_{biomass}}, \qquad (3.60)$$

are calculated based on the market sizes ms_{chem} and $ms_{biomass}$, the output fluxes of the chemicals b_{chem} or the biomass supply flux f_{S1}, respectively. For the value-added chemicals, access to the global market is assumed, while the biomass market is restricted to the local borders of the supply chain design space, without any in- or export of biomass.

By means of the introduced price functions, the plant's total annual revenues TR,

$$TR = \sum_{chem=1}^{N_{chem}} P_{chem} b_{chem} M_{chem} + \sum_{fuel=1}^{N_{fuel}} P_{fuel} \frac{b_{fuel} M_{fuel}}{\rho_{fuel}}, \qquad (3.61)$$

can be determined based on the output flux $\mathbf{b}$, the molar mass, and in case of fuel production, the density ρ_{fuel}. Due to a variable price for chemicals P_{chem}, this equation is bilinear. A plant's total profit, TP, is then calculated based on the difference between total annualized revenues and cost, as well as biomass transportation cost

$$TP = TR - TAC - TBTC. \qquad (3.62)$$

In order to prevent an overproduction of chemicals, the global market demand for every chemical is set as an upper limit (Kim et al., 2011a; Cambero et al., 2016):

$$b_{chem} \leq ms_{chem}. \qquad (3.63)$$

A similar constraint is given in the supply chain design for the biomass market to prevent a biomass collection larger than the available capacity at each origin (Eq. 3.50).

3.5 Optimization problem formulation

In this section, the formulations of the optimization problems are presented. First, the formulation of the RNFA problem is discussed and mandatory adaptations are outlined in Section 3.5.1. Afterwards, the optimization problems for the analysis of processing pathways only (cf. Section 3.5.2) as well as of full value chains are described (cf. Section 3.5.3).

3.5.1 RNFA problem

Compared to the original RNFA formulation (cf. Eq. 2.9), the TAC are altered such that waste disposal cost are included, as described in Equation 3.20. Since the energy demand cannot be examined using the RNFA, the utility cost are zero at the RNFA stage. In order to allow a fair comparison between RNFA and PNFA, the same IC correlation (Eq. 3.37) is applied in this thesis. The adapted optimization problem is formulated as

$$\min_{f,b,y} \left\{ \begin{array}{c} TAC \\ EI \end{array} \right\}$$

$$\text{s.t.} \quad \mathbf{A} \cdot \mathbf{f} = \mathbf{b},$$
$$yield\ constraint\ (Eq.\ 2.2),$$
$$biomass\ composition\ (Eqn.\ 2.3 - 2.6),$$
$$TAC\ calculation\ (Eqn.\ 3.20 - 3.43), \qquad (3.64)$$
$$EI'\ calculation\ (Eqn.\ 3.44),$$
$$\sum_{i=1}^{N_{products}} b_{product,i} \Delta h_{comb,product} = \alpha,$$
$$\mathbf{f}, \mathbf{b} \geq \mathbf{0},$$
$$\mathbf{y} \in \{0,1\}.$$

which is a MINLP problem. Binary integer decisions are coupled to the reaction fluxes determining the number of active reaction fluxes as input variable for the IC function. The nonlinearity arises from the investment costs calculation only due to a multiplication of the NFU and capacity plus the exponent $Inv2$ (Eq. 3.37).

3.5 Optimization problem formulation

Capacity functions are often reformulated using a piece-wise linear approximation (You and Wang, 2011; Schwaderer, 2012; Marvin et al., 2013a; Sharifzadeh et al., 2015) allowing an implementation as mixed integer linear programming (MILP) problem. While MILP formulations require less computational time compared to MINLP problems, accuracy is higher for the latter. The introduction of the NFU into the IC function increases the nonlinearity, such that an approximation thereof is not straightforward. However, at RNFA level, the overall problem complexity is small such that solvability is not an issue. Therefore, a reformulation of the IC function is not required and only of theoretical interest.

To convert the bi-objective optimization problem into a single objective optimization problem, the ϵ-constraint method is applied (Miettinen, 1998). By expressing one of the two objective function as a constraint, Pareto optimal curves can be formed. Pareto optimality refers to a state where one objective cannot be further improved without the other objective being impaired. Hence, the curve depicts the trade-off between the two objectives allowing the decision makers to choose a process according to their preferences (Miettinen, 1998; Rangaiah, 2009).

In the RNFA, the environmental impact is discretized and applied as constraint for the minimization of the TAC. Alternatively, the Pareto front could also be determined rigorously using parametric optimization (e.g., Pertsinidis, 1992; Pertsinidis et al., 1998).

3.5.2 PNFA for processing pathways

For the PNFA, the objective functions applied are the economic efficiency in terms of TAC and the process energy demand (CED_{fuel}). The biomass and yield constraints are formulated similar to the RNFA. The extended version of the mole balance (Equation 3.1) is applied which considers additional fluxes for mixing and separation, the yield constraint is extended to cover these fluxes as well (cf. Section 3.1.1). The assessment of CED_{fuel} considers a simultaneous heat integration, which increase the complexity of the optimization problem. A discussion of a simultaneous versus a sequential consideration of heat integration is conducted.

Since it is in principle possible to use additional and excess biomass (components) for combustion in order to reduce the energy demand, CED_{fuel} is constrained to prevent an infinite combustion. The primary goal of biomass utilization is not the production of grid energy, thus, the lower boundary of CED_{fuel} is set to zero which allows a maximal energy supply equivalent to the process energy demand.

3 Process Network Flux Analysis

The bi-objective optimization problem can then be written as

$$\min_{f,b,y} \left\{ \begin{array}{c} TAC \\ CED_{fuel} \end{array} \right\}$$

s.t. $\mathbf{A} \cdot \mathbf{f} = \mathbf{b}$,

$yield\ constraint\ (Eq.\ 2.2)$,

$biomass\ composition\ (Eqn.\ 2.3 - 2.6)$,

$TAC\ calculation\ (Eqn.\ 3.20 - 3.43)$,

$CED_{fuel}\ calculation\ (Eqn.\ 3.2 - 3.17, 3.19)$, (3.65)

$$\sum_{i=1}^{N_{products}} b_{product,i} \Delta h_{comb,product} = \alpha,$$

$CED_{fuel} \geq 0$,

$\mathbf{f}, \mathbf{b} \geq 0$,

$\mathbf{y} \in \{0,1\}$.

The optimization problem is again formulated as an MINLP problem. Compared to the RNFA, a higher number of binary variables exist, as further binary variables are required, indicating whether the separation, boiler and turbine fluxes are activated. The nonlinearity originates from the IC calculation as well as from the split between steam and electricity generation (Eqn. 3.9-3.10). Similar to the RNFA, the bi-objective problem is converted into a single objective problem by means of the ϵ-constraint method (Miettinen, 1998) discretizing now CED_{fuel}. While specialized solution methods are conceivable to solve this optimization problem, the more efficient general-purpose solver BARON by Tawarmalani and Sahinidis (2005) is used.

3.5.3 PNFA for value chains

The PNFA for processing pathways can be extended by a biomass supply chain design under the restriction of the biomass market, which has been outlined in Sections 3.3 and 3.4. To incorporate the influence of transportation on the overall sustainability, the global warming potential is now addressed as an environmental objective function, combining the process energy demand and the transportation into one single metric. This leads to an extended bilinear equation for the GWP determination:

$$GWP = GWP_{process} + \frac{GWP_{BT}}{\sum_{i=1}^{N_{products}} b_{product,i} \Delta h_{comb,product}}.$$ (3.66)

3.5 Optimization problem formulation

Furthermore, the analysis is extended, covering now the market model for biomass and value-added products. This includes a change of the economic objective function to the maximization of the profit. Compared to the original PNFA for process design, the problem complexity is increased due to the additional supply chain model, as well as the market and profitability analysis.

The nonlinear terms in the resulting mixed-integer nonlinear programming problem are mainly bilinearities, which are present in Eqn. (3.20), (3.19), (3.45), (3.61) and (3.66). While the general flux balance is linear, the nonlinearities are introduced for computation of the objective functions, namely the economic performance in terms of the profit and the global warming potential, which is nonlinear for a multi-product analysis.

Compared to the optimization problem for processing pathways only (Eqn. 3.65), additional binary variables are introduced for every potential plant site. In addition to the process fluxes, the biomass transportation fluxes increases the number of variables. Since the biomass transportation fluxes correlate with the number of origins, hence with the considered grid sizes, a comparison of the CPU time and the accuracy of the results is conducted later in Chapter 5. The optimization problem is formulated as:

$$
\min_{f,b,y} \left\{ \begin{array}{c} -TP \\ GWP \end{array} \right\}
$$

$$
\begin{aligned}
\text{s.t.} \quad & \boldsymbol{A} \cdot \boldsymbol{f} = \boldsymbol{b}, \\
& yield\ constraint, \\
& biomass\ composition, \\
& profit\ calculation\ (Eqn.\ 3.20, 3.54-3.55, 3.61-3.62), \\
& GWP\ calculation\ (Eqn.\ 3.45, 3.56, 3.66), \\
& supply\ chain\ design\ (Eqn.\ 3.50-3.53) \\
& biomass\ price\ constraints\ (Eqn.\ 3.58-3.60) \\
& market\ and\ price\ constraints\ (Eqn.\ 3.57-3.63) \\
& \sum_{i=1}^{N_{fuels}} b_{fuel,i} \Delta h_{comb,fuel} \geq \alpha, \\
& CED_{fuel} \geq 0, \\
& \lambda = 1, \\
& \boldsymbol{f}, \boldsymbol{b} \geq \boldsymbol{0}, \\
& \boldsymbol{y} \in \{0,1\}.
\end{aligned}
\qquad (3.67)
$$

3.6 Summary and conclusions

In this chapter, the PNFA is introduced as an extension to the existing RNFA methodology. The PNFA is an early stage design method for the evaluation of process pathways as well as for the identification of bottlenecks. It is not based on reaction pathways only, but rather benchmarks the process performance, considering in addition to the RNFA, the influence of separations, their feasibility, energy and solvent demands. As the choice of purification strategy strongly influences the process, a systematic evaluation of alternatives based on thermodynamic insights, i.e., pure-component property ratios determining the thermodynamic driving force of a separation, is conducted. Different separation strategies can be examined in the optimization by separate fluxes thereof. The process performance is assessed based on thermodynamically sound separation methods, which have the advantage of being independent of design decisions, e.g, the number of column stages. This circumvents tedious literature searches or simulation studies, which typically lack robustness and require many design decisions. This enables a fast and reliable identification of the most promising pathway.

In the PNFA, the pathways are evaluated based on their economic efficiency and sustainability for which a large number of criteria are presented and their applicability and significance discussed. Furthermore, process integration techniques like heat integration can be considered during optimization. In order to analyze a full biorefinery value chain from biomass transportation to the final products, PNFA takes the biomass supply chain as well as a pragmatic market model into account. Therefore, biomass availability and transportation are considered along with the identification of an optimal plant location and product-portfolio selection in dependency of the products' market exploitation. Bi-objective optimization problems are set up to analyze the trade off between economics and sustainability. The PNFA is coupled to a sensitivity analysis to address parametric uncertainties. In the next chapter the validity of PNFA results is proven before the PNFA is used to analyze novel process pathways in Chapter 5.

Chapter 4

PNFA benchmarking with literature, RNFA and conceptual design results

In order to provide a vision on the usefulness of the PNFA, it is mandatory to compare the accuracy of the PNFA results with literature, RNFA and conceptual design results. However, a fair comparison of PNFA results with available processes from literature is difficult, as the PNFA aims at the evaluation of novel processing concepts, for which a detailed design is often not available. Furthermore, the process performance strongly relies on design decisions, case study specific process parameters, feedstock and technology choices, which renders a detailed comparison challenging. Nevertheless, for processes available in literature, a comparison of the PNFA results is conducted in Section 4.1. For this purpose, the minimum selling prices of ethanol, iso-butanol and ethyllevulinate are discussed. Since further information on the investment cost and the global warming potential are not available for the latter two, these results are only discussed for ethanol production. The mandatory PNFA results used here for comparison are presented in more detail in Chapter 5.

The PNFA is embedded in a framework to evaluate novel bio-processes (cf. Section 2.3) such that the advancement from the RNFA to the PNFA and to conceptual design is discussed in Section 4.2. For this purpose, a conceptual design is conducted. Herein, ethanol production is chosen as most prominent biorefinery process for which a vast amount of data including reaction kinetics are available. In addition, the conceptual design is also used to overcome the difficulties of a fair literature comparison and thus enables a detailed comparison of the PNFA screening results with a rigorous model.

4 PNFA benchmarking with literature, RNFA and conceptual design results

4.1 Cost and GWP comparison of PNFA and literature

The following comparison is conducted in order to prove the reliability of the PNFA results. For this purpose, the PNFA results of Section 5.1.7.2 are utilized for the comparison, which include contributions from biomass transportation and processing. In case of discrepancies, a brief discussion is provided. All available literature studies for the analyzed processes are considered for the following comparison. While often minimum selling prices are published, a breakdown of the costs into individual cost contributions is missing, which impedes a comparison of the IC in Section 4.1.2.

4.1.1 Comparison of minimum selling prices

Until now, the most studied and mature bio-process to date is ethanol production. It is taken to assess the results by PNFA in comparison with literature. According to Balat (2011), the ethanol production cost range is between 0.2 and 1.1 $/L, but is expected to decrease to 0.2-0.65 $/L. This is in line with the references listed in Table 4.1, which report MSPs for ethanol between 0.29 and 1.29 $/L. Table 4.1 also briefly summarizes the plant's capacity, type of feedstock and lists all cost contributions considered in the individual references. Furthermore, studies which optimize the plant capacity are marked specifically. The comparison of the various references illustrates the variety of results depending on the feedstock, the chosen processing type, considered cost factors and selected cost parameters. In addition, these references either work with rigorous models or utilize data of detailed process designs.

The production of ethyllevulinate further illustrates the challenges of comparing PNFA with literature results. Hayes et al. (2006) describe a plant converting lignocellulosic biomass to ethyllevulinate with an annual production output of 133,000 tons. Even though the annual plant capacity is similar to the capacity analyzed in the PNFA (111,406 tons/year), a lower minimum selling price (MSP) of 0.3 $/L (Hayes et al., 2006) compared to 0.75 $/L (PNFA) is proposed. Three major reasons explain the discrepancies between the results. First, the raw material price utilized by Hayes et al. (2006) is 20% cheaper. Second, Hayes et al. (2006) consider revenues from byproducts, which naturally lower the MSP. Third and most importantly, the MSP of Hayes et al. (2006) neither includes a contribution from investment nor from transportation. If all these factors are adapted, the PNFA result is lowered to 0.27 $/L, which agrees well with the result of Hayes et al. (2006). The remaining small discrepancy is caused by a different process structure and, thus, likely different yields, which are not disclosed by Hayes et al. (2006) for all process steps.

Table 4.1: Minimum ethanol production cost.

Reference	MSP $/L	capacity Mio. L	description
PNFA	0.39	126	feedstock: lignocellulosic biomass cost: IC, raw materials, utilities, transport
Huang et al. (2010)	0.29	51[1]	feedstock: biowaste cost: IC, raw materials, utilities, transport, procurement
Ekşioğlu et al. (2009)	0.45	219[1]	feedstock: lignocellulosic biomass cost: IC, raw materials, utilities, transport, feedstock collection, inventory
Marvin et al. (2013b)	0.47	44	feedstock: lignocellulosic biomass costs: IC, raw materials, utilities, transport, taxes, maintenance
Tao et al. (2014)	0.57	230	feedstock: lignocellulosic biomass cost: IC, raw materials, utilities, transport, feedstock collection, inventory
Humbird et al. (2012)	0.57	230	feedstock: corn stover cost: IC, raw materials, utilities, waste disposal, salaries, taxes
Archambault-Léger et al. (2015)	0.82	224	feedstock: sugarcane bagasse cost: IC, raw materials, utilities, waste disposal
You et al. (2012)	0.85	6 plants[1]	feedstock: agricultural residues, energy crops, wood residues cost: IC, raw materials, utilities, transport, storage and handling individual capacities: 567 (4 plants), 564, 556 million L/year
You et al. (2012)	0.97	4 plants[1]	feedstock: corn, agricultural residues cost: IC, raw materials, utilities, transport, storage and handling individual capacities: 386, 469, 522, 567 million L/year
Klein-Marcuschamer et al. (2010)	1.21	127	feedstock: corn stover cost: IC, raw materials, utilities
Luterbacher et al. (2014)	1.29	230	feedstock: corn stover cost: IC, raw materials, utilities, waste disposal

[1] The plant capacity is an optimization variable.

Another example is the iso-butanol production for which the PNFA determines a MSP of 0.57 $/L. Tao et al. (2014) describe a price of 0.78 $/L and Gevo (2016) a company selling price of 1.19 $/L. The latter value is a company perspective, thus, it is likely that it includes taxes, wages and revenues in addition to the investment and operating cost. Therefore, this value presents an upper limit, as the described factors (taxes, wages, company revenues) are not included in the PNFA. Similarly, the value of Tao et al. (2014) also includes factors (fixed cost for labor, supply and overheads, taxes, average return on investment), which are not considered in the PNFA, thus explaining a lower MSP in the PNFA. A more detailed comparison is not possible, as the cost structure and the respective price parameters are not fully disclosed by Tao et al. (2014). Interestingly, the gap between the MSP of ethanol and iso-butanol using the PNFA is similar to the gap described by Tao et al. (2014).

More importantly, the examples illustrate the difficulties of a fair process comparison. The reasons are (i) different feedstock selections and moisture contents of the biomass, (ii) different cost factors, which are considered in the cost analysis, (iii) variations in the price parameters and (iv) plant capacities as well as (v) differences in the level of modelling detail. Screening results are compared to rigorous models and even the rigorous models do not all consider process integration techniques.

As the example comparison of the ethyllevulinate production shows, the first four aspects are more significant, as a PNFA adaptation of the price parameters and cost factors levels the differences to a literature value out. However, often not the full set of parameters and required data is disclosed in literature, such that a simple and fair process comparison is impossible. The comparison also proves a validation based on the MSP alone inefficient. Instead, the comparison always needs to be accompanied by a description of the considered cost contributions and plant capacities. The latter is of special importance to biorefineries, since higher relative investment cost contribution to the overall costs are observed (cf. Section 5.1.4). Nevertheless, the PNFA value is within the MSP ranges for all three examples, which demonstrates the reliability of the PNFA results.

4.1.2 Comparison of investment costs

A fair IC comparison is difficult, since the accuracy depends on the stage of process development and, thereby, process knowledge. When only basic knowledge is available, such as the capacity, number of steps or process conditions, uncertainty of up to 100% may be present (Tsagkari et al., 2016). The order-of-magnitude correlations are sufficient for a first flowsheet screening (Douglas, 2000) and are therefore applied.

4.1 Cost and GWP comparison of PNFA and literature

More detailed unit-wise cost correlations, which consider additionally, operating conditions, choice of material and a unit operation characteristic size, like heat exchange area, still have uncertainties between 25 and 40% (Biegler et al., 1997). Furthermore, the IC strongly correlates with the process design and structure as well as with the biomass feed and pretreatment technology. Since the ethanol production is a prominent biorefinery process, data and IC are available in literature and are listed in Table 4.2.

Table 4.2: Overview of IC for an ethanol production plant of various capacities, which are scaled to an annual reference production of 100,000 t ethanol.

reference	$Cap_{literature}$ $\frac{t\ ethanol}{year}$	$IC_{literature}$ Mio. $	$IC_{reference}$ Mio. $
PNFA	100,000	-	147-165
Tao et al. (2014)	134,100	232	190
Klein-Marcuschamer et al. (2010)	100,000	340	340
Archambault-Léger et al. (2015)	176,736	375	255
Humbird et al. (2012)	182,188	400	266

To enable an easier comparison, the IC from literature $IC_{literature}$ are scaled to the ethanol reference production with a capacity $Cap_{reference}$ of 100,000 $\frac{t\ ethanol}{year}$, according to

$$IC_{reference} = IC_{literature}(\frac{Cap_{reference}}{Cap_{literature}})^{Inv2}. \qquad (4.1)$$

Values for $Inv2$ typically range between 0.6 and 0.7 (Humbird et al., 2012). Thus, the value of 0.68 is taken from the PNFA for all cases. This leads to IC of the reference production $IC_{reference}$ between 190 and 340 Mio. $. These wide ranges demonstrate the high uncertainties even for detailed process designs. The comparison of the literature values to the PNFA results of 147 Mio. $ for a process excluding and 165 Mio. $ for a process including combustion, demonstrates that the IC are 13% - 57% lower compared to literature. The deviation of 57% originates from the comparison with the values of Klein-Marcuschamer et al. (2010), which are nearly twice as high compared to the values of Tao et al. (2014). The literature values for the detailed design are far apart from each other, such that the IC obtained with the PNFA are sufficiently accurate for the early design stage since the IC are in in the uncertainty range described by Tsagkari et al. (2016) and in the right order of magnitude.

4.1.3 Comparison of global warming potential

A GWP comparison is even more challenging than the cost comparison, as often a detailed description of individual GWP contributors is missing. Large GWP differences up to 70 $\frac{\text{g CO2}}{\text{MJ}}$ occur depending on the feedstock and nearly 15 $\frac{\text{g CO2}}{\text{MJ}}$ depending on geographical differences for otherwise similar processes (Benalcázar et al., 2017). Tao et al. (2014) propose values between 49 and 127 $\frac{\text{g CO2}}{\text{MJ}}$ and You et al. (2012) a value of 311 $\frac{\text{g CO2}}{\text{MJ}}$ for an ethanol production. This demonstrates that the ranges described in literature vary substantially.

Herein, the PNFA considers only non-renewable greenhouse gas emissions caused by utilities and transportation. For most ethanol biorefineries, a self-sufficient production is described such that no external energy and only few external utilities are required (Humbird et al., 2012). This is in line with the values obtained in the PNFA for a scenario including residue combustion for which only a small GWP contribution results from transportation.

When excluding residue combustion, the PNFA determines a GWP of 16 $\frac{\text{g CO2}}{\text{MJ ethanol}}$. This PNFA value is similar to the value of nearly 20 $\frac{\text{g CO2}}{\text{MJ}}$ for a biochemical conversion of woody biomass to ethanol considering transportation, production and product distribution (Edwards et al., 2014). In conclusion, the comparison of PNFA results with literature demonstrate the validity of GWP results for bio-ethanol production.

4.2 Accuracy of the PNFA results for an ethanol production compared to RNFA and conceptual design

The target of this section is to compare the accuracy of the RNFA, PNFA and conceptual design in order to demonstrate the framework presented in Section 2.3. For this purpose, a bio-ethanol production is described. Herein, ethanol is chosen as required data for the conceptual design, e.g., reaction kinetics, are available in literature.

For the RNFA and PNFA the optimization problems described in Equation 3.64 and 3.65 are solved, respectively. A short model summary of the conceptual design is given in Section 4.2.1. Section 4.2.2 describes the optimization results thereof and Section 4.2.3 then compares the results of the RNFA, PNFA and the conceptual design.

4.2.1 Conceptual process design for ethanol production

A multitude of different processing concepts are conceivable for the production of bio-ethanol. The general processing steps are classified into biomass pretreatment, enzymatic hydrolysis, fermentation and downstream processing. Each step is briefly summarized. The reaction kinetics and modelling approaches used, are summarized in Table C.24 of Appendix C.10.

The goal of biomass pretreatment is a biomass split into its main constituents, a reduction of cellulose crystallinity, a (partial) hydrolysis of the hemicellulose fraction as well as the prevention of sugar degradation and a production of toxic by-products (Mergner et al., 2013). A dilute-acid technology for pretreatment is considered, as it is the most prominent pretreatment concept. Herein, sulphuric acid is utilized as an efficient and cheap chemical (Kumar et al., 2009). In the pretreatment step, lignin is partially separated and a large hemicellulose fraction is hydrolyzed and diluted. Neutralization is achieved with ammonia. Along with sulphuric acid, ammonia forms ammonium sulphate, which later serves as nutrient in the fermentation and, thus, does not need to be separated.

In case of a separate hydrolysis and co-fermentation (SHCF), the remaining solids are separated after the enzymatic hydrolysis. In case of a simultaneous saccharification and co-fermentation (SSCF) a separation of solids is not favorable since cellulose is still present as solid. Thus, the separation of solids is conducted after the SSCF. The liquid product stream of the enzymatic hydrolysis is used in the fermentation, which converts the sugars into ethanol. The microorganisms during fermentation are mainly influenced by the choice of pH and temperature, nutrient supply and the presence of inhibitors. A co-fermentation allows a simultaneous conversion of pentoses and hexoses. For the fermentation, the well studied *Saccharomyces cerevisiae* is utilized, which is an inhibitor-resistant microorganism (Morales-Rodriguez et al., 2011). A solid-liquid separator is placed after the fermentation to separate the microorganisms from the product stream.

The liquid product stream is then purified in the downstream processing, for which a multitude of alternatives exist. Examples are distillation with subsequent purification using molecular sieves, extractive distillation, membrane-assisted vapor stripping (Vane, 2008) or distillation coupled to pervaporation (Skiborowski, 2014). Even though, the combination of distillation and molecular sieves is wide-spread, an energy demand comparison identifies the setup consisting of distillation and pervaporation as favorable (Skiborowski, 2014). Thus, ethanol concentration is increased in a beer column and further dehydrated using a hybrid process of pervaporation and distillation.

4 PNFA benchmarking with literature, RNFA and conceptual design results

The overall process structure is summarized in Figure 4.1.

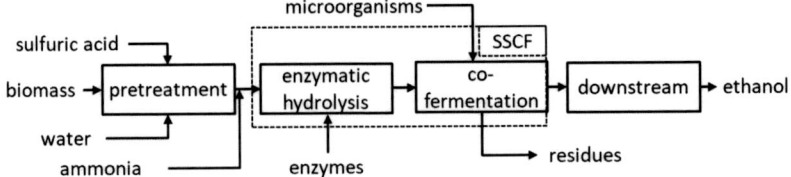

Figure 4.1: General process structure for bioethanol production.

The process structure is optimized to minimize the specific ethanol production cost, which consists of the raw material and annualized investment costs. Instead of a rough order of magnitude IC estimation as applied in the PNFA, a cost calculation based on unit-wise cost estimation according to Guthrie (1969) is conducted. Therefore, a more precise IC determination is enabled, which serves as benchmark for the IC determined with the PNFA. All price parameters and cost equations are given in Appendix C.10. A biomass feed of 100 $\frac{t}{h}$ is set as a design constraint.

The resulting optimization problem is a nonlinear programming problem. The nonlinearity mainly arises from the reaction rates and the cost determination. Since no recycles from downstream to upstream processing are considered, the models are not connected. Therefore, the nonlinearity resulting from thermodynamic models is not important. The process is modelled in GAMS version 24.6.1 (GAMS Development Corporation, 2012). As a consequence of the high nonlinearities, an elaborate initialization procedure is required. To determine suitable starting parameters, the mass balances of each processing step are solved. Subsequently, a yield optimization for the desired product of the specific processing step is conducted. The resulting outlet stream is then fixed as the inlet stream of the following process step. For these first steps, SNOPT is applied as solver since it is capable of efficiently determining a local solution based on a few parameters and starting value specifications. The optimization of the full process is then conducted using CONOPT. The solvers BARON and ANTIGONE have been tried as well to obtain a global solution. An upper time limit of 72 hours has been set. BARON did not converge within this time limit. The lower bound remained close to zero, whereas the upper bound was similar to the local solution obtained with CONOPT. ANTIGONE converged after four hours and found the same local solution, which was obtained with CONOPT. The following section presents the optimization results of the conceptual design before comparing them to the results obtained with the PNFA in Section 4.2.3.

4.2.2 Conceptual design results

The minimization of the specific ethanol production cost reveals a marginally superior performance of the SSCF concept compared to the SHCF structure. Even though the overall process yield with the SSCF is 0.219 $\frac{\text{kg ethanol}}{\text{kg biomass}}$, which is lower than the yield of the SHCF (0.222 $\frac{\text{kg ethanol}}{\text{kg biomass}}$), the minimum production cost for the SHCF is 0.65 $\frac{\text{€}}{\text{L ethanol}}$ and for the SSCF 0.63 $\frac{\text{€}}{\text{L ethanol}}$. Thus, only the SSCF results are presented in more detail.

For biomass pretreatment a plug-flow reactor with a reactor volume of 22.74 m^3 and a residence time of 4.08 minutes is determined to be optimal. With only 35.8% of the theoretical yield, the xylose yield is the major bottleneck. The reason is an increased degradation at high xylose concentrations. A longer residence time would further increase the degradation and a single continuous stirred tank reactor (CSTR) leads to a xylose yield of only 21.9% of the theoretical maximum. Furthermore, the kinetic model of Lavarack et al. (2002) lacks accurate parameters at higher temperatures. Therefore, a model validation is required for a future improvement.

While the biomass pretreatment is similar for both subsequent process concepts (SSCF, SHCF), the optimal operating point of the SSCF reveals a longer residence time for the enzymatic hydrolysis and fermentation compared to the SHCF. This explains the lower process yield as more substrate is required over a longer time as nutrient source to maintain metabolism. In addition, cellulose is not fully hydrolyzed and the enzyme loading is at its lower bound. A total reactor volume of 7604 m^3 along with a residence time of 22.6 h is determined. Product inhibition in the enzymatic hydrolysis is circumvented in the SSCF due to direct consumption of sugars by the microorganisms.

The feed and product mass streams are shown in Figure 4.2. Different reasons for carbon losses are identified. Lignin is not converted and thus fully lost. Parts of the hemicellulose fraction are not hydrolyzed due to the short residence time during pretreatment and simultaneously a minor pentose fraction is degraded. In the subsequent SSCF, carbon losses occur due to the stoichiometric production of carbon dioxide. In addition, cellulose is not fully converted and carbon is required as nutrient for metabolism. The splitter separates the solids from the liquids. As a sharp split is not achieved, caused by a residual moisture content in the filter cake, small losses of ethanol occur. Water recycling is not considered in the process as impurities influence the microorganisms' performance. In order to reduce the amount of waste water, this needs to be analyzed in more detail in future.

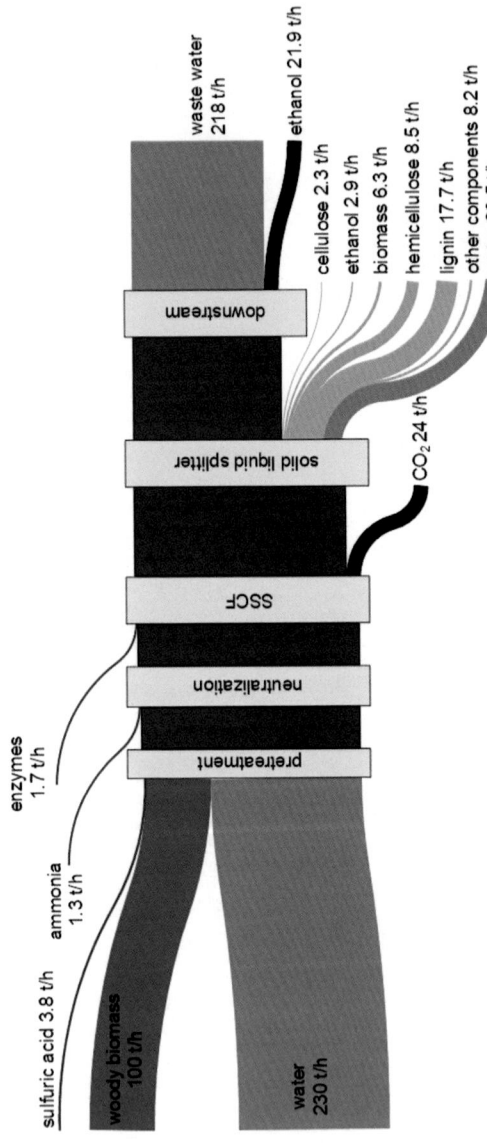

Figure 4.2: Mass flows at the optimal operating point of the process structure containing the simultaneous saccharification and cofermentation (SSCF).

Major costs are the biomass raw material cost (48.9%) and the enzyme cost (36.2%). Auxiliary costs for ammonia, sulfuric acid and water are low with a cost contribution of 2.6%. The annualized investment cost are responsible for 11.4% and the costs for downstream processing for 0.9% of the minimal ethanol production costs. The high cost contribution of the enzymes is in line with literature. In literature, enzyme costs contribution up to 25 cent per liter are postulated for lignocellulosic ethanol (Graham-Rowe, 2011; Klein-Marcuschamer and Blanch, 2015; Brodeur et al., 2011), and 2 cent for corn ethanol (Graham-Rowe, 2011). To reduce the enzyme cost, a further reduction of the enzyme loading is targeted (Brodeur et al., 2011). This is in agreement with the optimization results, where the enzyme loading is at its lower bound, independent of the process structure (SSCF, SHCF). Besides enzyme loading, which affects enzyme cost, another major bottleneck is the pentose yield during pretreatment.

4.2.3 Comparison of RNFA, PNFA and conceptual design results

In the following, the results of an ethanol production are compared using the RNFA, PNFA and the conceptual design. The model characteristics are once more summarized in Table C.25 of Appendix C.10. The results for all three methods are presented in Table 4.3. Overall, the RNFA determines the lowest ethanol production cost with 0.33 $\frac{€}{L}$. The highest production cost of 0.63 $\frac{€}{L}$ are obtained using the rigorous model. The production cost determined with the PNFA are in between with 0.41 $\frac{€}{L}$. Thus, the results reflect the accuracy level of the methods. With an increasing model complexity, the investment cost raise as more cost factors are considered.

Table 4.3: Specific ethanol production cost.

Cost factors	RNFA [$\frac{€}{L}$]	PNFA [$\frac{€}{L}$]	Rigorous model [$\frac{€}{L}$]
Annualized investment	0.12	0.17	0.07
Biomass	0.21	0.21	0.31
Enzymes	-	-	0.23
Auxiliaries	0	0.03	0.03
Sum	0.33	0.41	0.63

The IC of the RNFA is lower compared with the PNFA. The reason is that separation steps are not included in the RNFA and therefore, the number of functional units is lower, which leads to lower IC. The IC of the PNFA is higher compared with the detailed model, but for a first rough estimation the accuracy is adequate.

The pretreatment yield is lower than that assumed in the RNFA and PNFA. This causes lower raw material cost with the RNFA and PNFA compared to the detailed model. Since enzymatic hydrolysis is only assessed using a yield parameter in the screening, no enzyme cost are determined. Both aspects (pretreatment, enzyme cost) effect all processes in a similar manner and, thus, do not influence a process ranking.

The RNFA underestimates the auxiliary cost, since no energy demand is determined in the model. In contrast, similar auxiliary cost are determined for the PNFA and the detailed model. Therefore, only a low discrepancy of the process energy demand between the PNFA and the detailed model is observed. The detailed model determines a process energy demand of 0.3 $\frac{MJ}{MJ_{fuel}}$ and the PNFA of 0.22 $\frac{MJ}{MJ_{fuel}}$. The reason for this discrepancy is again the biomass pretreatment, as an energy demand for pretreatment is not yet included in the PNFA. If residue combustion for steam production is considered, both models determine a self-sufficient production plant.

4.3 Summary and conclusions

In this chapter, PNFA results are compared to literature, RNFA and conceptual design results. The comparison of the PNFA results to literature data is difficult, as the results strongly correlate with the feedstock selection, design and technology choices, the cost contributions considered in the economic analysis, case study specific process parameters, plant capacities and the level of modelling detail.

Furthermore, often not all required data and process parameters are disclosed in literature, such that a detailed comparison is challenging. The minimum selling prices of ethyllevulinate, iso-butanol and ethanol production fit to the literature ranges.

Wide IC and GWP ranges exist in literature for ethanol production. The investment cost obtained with the PNFA is lower compared to the literature studies, but in the right order of magnitude. The global warming potential obtained using the PNFA fit well to a literature study for a woody ethanol process concept.

The comparison of the RNFA, PNFA and conceptual design results for an ethanol production demonstrate the methodological advancement of the PNFA compared to the RNFA. Major difference is the consideration of the process energy demand. Compared with conceptual design, the energy demand determined with the PNFA leads to similar results.

The ethanol production cost increase from RNFA to PNFA and conceptual design. This can be explained by an increase in the level of detail and process knowledge.

4.3 Summary and conclusions

Due to xylose degradation effects and a limited temperature validity range of the kinetic model for pretreatment, the process yield is lower in the conceptual design compared with the RNFA or PNFA. This causes higher raw material costs. Furthermore, enzyme cost are neither considered in the RNFA nor in the PNFA. An introduction of enzyme costs rises the production cost, such that the cost obtained using the PNFA sets a lower bound on the production cost.

The differences between the PNFA and conceptual design effect all processes in a similar manner. Therefore, the process ranking is not influenced by these factors. Furthermore, the example demonstrates how the three stages (RNFA, PNFA and conceptual design) of the biorefinery process framework build up on each other in order to evaluate a process performance of novel pathways.

Chapter 5

Application of PNFA

This chapter presents case studies demonstrating the applicability and potential of the novel PNFA methodology. The findings are compared to results from the previously existing screening methodology RNFA to outline the differences between both approaches and to emphasize the knowledge gain from the extension to PNFA. The major goal is the analysis of novel process concepts to obtain viable processes. For this purpose, important aspects like heat integration, biomass transportation and co-production are discussed.

In Section 5.1, a complex case study for six biomass-derived gasoline fuels is presented, which covers in depth all aspects of the PNFA. In order to obtain viable processes, fuel mixtures as well as co-production scenarios thereof are analyzed in Section 5.2. In Section 5.3, the influence of co-producing value-added chemicals on process viability is discussed.

The RNFA methodology has also been applied to screen a large number of biofuels containing alcohols, alkanes, ethers, levulinates, furans, tetrahydrofurans, lignin-based fuels and gaseous fuels. The results are published in Ulonska et al. (2016b). Both, the RNFA and PNFA methodology are as well utilized to screen and evaluate the process performance of biofuels in comparison with fuels derived using renewable electricity, which is published in König et al. (2018). Both case studies demonstrate the applicability of the RNFA and PNFA on a large case study, but the results are not shown herein.

5 Application of PNFA

5.1 Single-product biorefinery for fuel production

In the following, a case study for six gasoline biofuels is presented. In literature, the components ethanol, γ-valerolactone, ethyllevulinate, iso-butanol, 2-butanol and 2-butanone have been proposed as bio-based gasoline fuels. All of them have a Research Octane Number equal or larger than 100, which qualifies them as fuels for spark-ignition engines (Yanowitz et al., 2011; Thewes et al., 2012; Hoppe et al., 2015). All fuels are produced from lignocellulosic biomass. As none of the fuels can be produced from lignin, its utilization is not considered in the RNFA, while lignin combustion is used for internal energy supply in the PNFA. All information on the reactions, reaction and separation parameters, pure and mixture properties as well as active reactions and separations are given in Appendix C.1 - C.5.

Results of the RNFA and PNFA are shown as well as the knowledge gain from the methodological advancement. Both, the RNFA and PNFA are implemented in GAMS version 24.7 (GAMS Development Corporation, 2012) using BARON (Tawarmalani and Sahinidis, 2005) to solve the MINLP problem. BARON is chosen as solver to obtain global solutions. All calculations have been performed on an Intel(R) Core(TM) i5-6200 CPU (3.10 GHz) with a relative tolerance (optcr) of 0.01. The tolerance limit is therefore lower than the expected uncertainties in the methodologies or in the model parameters.

5.1.1 Reaction network

It is mandatory to first collect all the different reaction datasets. For this purpose, a reaction network is created using the online database Reaxys (Elsevier Information Systems GmbH, 2015) complemented by a literature search gathering all possible reactions from the biomass fractions cellulose, hemicellulose and lignin to the respective biofuels. In a second step, a pre-selection is required in case of multiple setups for the same reaction. In these cases, only the setup with the highest reported yield and bio-based solvent is considered.

The final reaction network is shown in Figure 5.1 consisting in total of 33 reactions. Lignin is not valorized herein, since a former RNFA study proved a selective conversion of lignin into fuels as too expensive even though an ideal and optimistic decomposition into its major constituents (coumarylalcohol, sinapylalcohol and coniferylalcohol) was assumed (Ulonska et al., 2016b). Thus, lignin is not valorized in the RNFA, but for an internal energy supply in the PNFA (cf. Section 5.1.4). In addition, more value-added products for the utilization and exploitation of lignin are targeted in the future.

5.1 Single-product biorefinery for fuel production

While most of the remaining reactions are performed using water as solvent, the reactions to ethyllevulinate (R10, R25, R26, R28) do not require an additional solvent and use an excess of the co-reactant ethanol instead. Few reactions need an additional solvent, like dimethylsulfoxide (R12), 2-methyltetrahydrofuran (R19), 1,4-dioxane (R27), methylenechloride (R37, R39) or γ-butyrolactone (R31, R33) which are supplied externally.

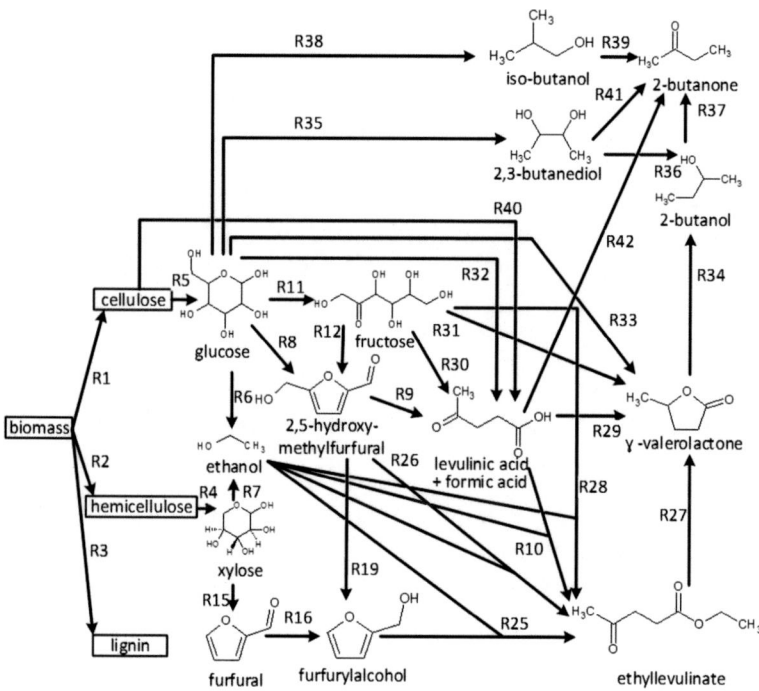

Figure 5.1: Reaction network for the production of gasoline fuels from lignocellulosic biomass.

5.1.2 RNFA results

The RNFA results are obtained by solving the optimization problem 3.64 which enables a direct comparison of the RNFA and the PNFA (Equation 3.65). The aims of the RNFA are (i) an evaluation of the reaction performance regarding cost and environmental impact, (ii) a ranking of processes to identify the best performing process and (iii) a reduction of the number of reaction alternatives to the most promising ones.

Figure 5.2 shows the Pareto curves for all six biofuels with the trade-offs between environmental impact and cost. If a single pathway is superior in terms of EI' and cost, the points on the Pareto curve coincide, such that only a single point is shown. This is the case for 2-butanone production (pathway R5, R35, R41). However, in the case of iso-butanol, there is only one point shown as only one possible pathway exists (R5, R38). In all other cases, multiple points exist which represent different pathways.

The relative ranking reveals the pathways to 2-butanol as worst performing, while the residual five products are within the same cost and EI' range. Only 2-butanone and iso-butanol exhibit a slightly higher EI', which can be explained by a higher resource consumption and a higher loss of the enthalpy of combustion caused by a lower yield of the pathways compared to ethanol. Reasons for the poor performance of 2-butanol are the low yield of reaction R36 and the low selectivity to 2-butanol in reaction R34 rendering the production of 2-butanol inefficient.

The analysis of the network for all six biofuels lowers the number from 33 possible to 27 active reactions. The reactions to 2-butanone (R37, R39, R42) are not active as well as reactions R19, R27 and R36 (cf. Appendix C.5). The main reason for the small reduction of the number of reaction alternatives, is the assumption of conversion limitation, which enables a high resource exploitation when minimizing the environmental impact. This leads to a high number of parallel and therefore active reactions. This is especially true for the pathways to ethyllevulinate, γ-valerolactone and 2-butanol. In contrast, minimizing the cost leads to short pathways starting, for all analyzed products, from cellulose only.

The case study shows that the RNFA is capable of detecting mass-related bottlenecks on a reaction level and of comparing the performance. However, a clear identification of the most promising process cannot be achieved. The assumption of a conversion limitation and ideal separations are the reasons for a high number of parallel pathways and therefore active reactions. Hence, a significant reduction of the reaction alternatives is not obtained. Furthermore, the EI' is a lumped factor, based on weighting factors, which are always subject to discussion (Banimostafa et al., 2012).

5.1 Single-product biorefinery for fuel production

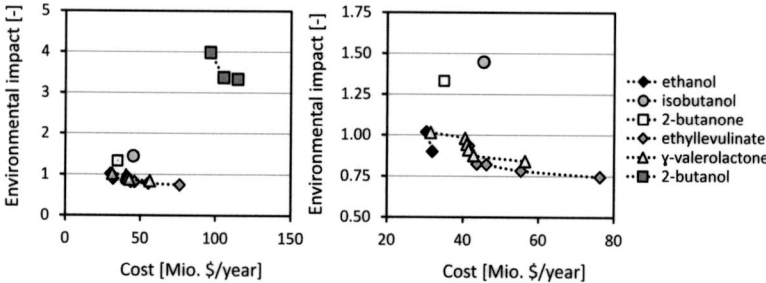

Figure 5.2: Left: RNFA results of all fuels shown as Pareto curves demonstrating the trade-off between environmental impact and cost based on a bi-objective optimization. Right: only the most promising fuels are presented as part of the left plot.

In the next step mandatory separations along with feasible separation techniques are identified for all reaction pathways.

5.1.3 Analysis of separations

In the following, the results of the feasibility analysis for all separations are discussed. For this purpose, Table 5.1 comprises all analyzed binary pairs, the associated reactions, information on azeotropes as well as the binary ratios of the boiling points T_B, the vapor pressure p_{oi}, the octanol-water partition coefficient $\log K_{OW}$, the molar volume V_M as well as the Hildebrandt solubility HS. Lower bounds are given for distillation, flash evaporation and extraction and their corresponding property ratios (Jaksland et al., 1995; Holtbruegge et al., 2014).

The feasibility analysis reveals several homogenous (VLE) and heterogenous (VLLE) azeotropes, while for the residual mixtures evaporation and zeotropic distillation are proven to be feasible based on the analysis of the boiling point and vapor pressure ratio. None of the binary ratios exceeds the lower bound for $\log K_{OW}$ rendering an efficient extraction unsuitable and even those mixtures with a $\log K_{OW}$ close to the lower bound of five (furfuryl alcohol + 2-methyltetrahydrofuran, γ-valerolactone + γ-butyrolactone, 2-butanone + methylenechlorid) have a molar volume or solubility ratio close to unity.

83

Table 5.1: Property ratios and azeotropes of all binary mixtures associated to the separation tasks and reactions.

Separation task	Reactions	Azeotrope	T_B	p_{oi}	$\log K_{ow}$	V_M	HS
Lower bound distillation (Jaksland et al., 1995)			1.01	1.05	-	-	-
Lower bound flash evaporation (Jaksland et al., 1995)			1.23	10	-	-	-
Lower bound extraction (Holtbruegge et al., 2014)			-	-	5	1.9	1.3
ethanol - water	6,7,10,28	VLE	1.06	2	(-0.3)	3.4	2.2
fructose - water	11	-	1.73	1801	(-3.5)	6.3	1.3
2,5-hydroxymethylfurfural - water	8,12	-	1.34	120	(0.02)	5.7	1.7
2,5-hydroxymethylfurfural - dimethylsulfoxid	12	-	1.08	3	-0.01	1.3	1.0
dimethylsulfoxid - water	12	-	1.24	40	(-1.4)	4.3	1.8
furfuryl alcohol - water	16	-	1.19	40	(0.3)	4.8	1.9
furfuryl alcohol - 2-methyltetrahydrofuran	19	-	1.25	160	4.5	1.1	1.3
levulinic acid - water	9,30,32,40	-	1.42	11024	(-0.3)	6.0	2.0
levulinic acid - formic acid	9,30,32,40	-	1.42	19779	1.9	2.9	1.0
formic acid - water	9,28,30,32,40	VLE	1.00	2	(-0.5)	2.1	1.9
ethyllevulinate - water	10	VLLE	1.28	127	(0.7)	8.0	2.4
ethyllevulinate - ethanol	10,25,26,28	-	1.36	318	-2.2	2.4	1.1
ethyllevulinate - formic acid	28	-	1.28	228	-1.2	3.9	1.2
formic acid - ethanol	28	-	1.06	1.4	1.8	0.6	1.1
γ-valerolactone - water	29,31,33	-	1.29	62	(-0.1)	5.3	2.2
γ-valerolactone - γ-butyrolactone	31,33	-	1.01	1.2	4.8	1.2	1.1
γ-butyrolactone - water	31,33	-	1.28	53	(4.4)	2.0	2.4
γ-valerolactone - ethanol	27	-	1.37	154	2.3	1.6	1.0
γ-valerolactone - 1,4-dioxane	27	-	1.28	99	2.0	1.1	1.0
ethanol - 1,4-dioxane	27	VLE	1.07	2	1.1	1.4	1.0
iso-butanol - water	38	VLLE	1.02	2	(0.8)	5.3	2.3
2,3-butanediol - water	35	-	1.29	670	(-0.9)	5.0	1.9
2-butanone - methylenechlorid	37, 39	-	1.13	5	4.3	1.4	1.0
2-butanone - water	42	VLLE	1.06	4	(0.3)	5.1	2.5
2-butanol - water	34, 36	VLLE	1.00	1.4	(0.7)	5.2	2.3

5.1 Single-product biorefinery for fuel production

For mixtures with water, a lower bound of zero is set as feasibility criteria for the $\log K_{OW}$ value (cf. Section 3.2.1.2). Therefore, all binary mixtures exhibiting a VLLE and the system γ-butyrolactone - water are suitable for extraction. As the latter system appears in a mixture with γ-valerolactone as well, an efficient separation of the lactones using extraction is not ensured, while for the VLLEs the miscibility gaps simplify separations, which leads to energy-efficient heteroazeotropic distillations without the use of an additional extracting agent. Therefore, evaporation and (azeotropic) distillation are applied, while neglecting extraction.

5.1.4 PNFA results

The resulting process network contains now 116 fluxes for supply, reaction, mixing and separation instead of the former 37 for only supply and reaction. This leads to a model with 467 equations, 572 continuous and 106 binary variables. None of the single optimization runs exceeded a CPU time of 21 seconds.

The different pathways are analyzed first neglecting residue combustion (Scenario I). Afterwards, the results are compared to the results with internal energy supply (Scenario II). Subsequently, the product ranking and the cost structure of the processes are discussed based on a literature comparison.

Scenario I: PNFA results without internal energy supply In the following, first the PNFA results are compared to those obtained by the RNFA before analyzing the products and their pathways in more detail. The PNFA reduces the number of active fluxes from 116 to 45 and the number of active reactions from 33 to 16. Compared to 27 active reactions in the RNFA, the PNFA reduces the number significantly. While the active reactions basically remain the same for ethanol, iso-butanol and 2-butanone, fewer reactions are active for the residual products in the PNFA (cf. Appendix C.5). The reason for this is a change in the sustainability objective function used, as the minimization of the environmental impact favors high resource exploitation, while a minimization of the CED leads to a low number of energy-demanding separations and, therefore, a lower number of active reactions.

Figure 5.3 presents the PNFA results for the six fuels and reveals a superior performance of ethanol in terms of cost and energy demand. Similar to the RNFA, the process to 2-butanol is not competitive due to the low yield and a relatively high number of mandatory reactions and separations. Besides ethanol, iso-butanol can be identified as a promising alternative.

5 Application of PNFA

In the following, the processes are described in detail. The product is marked in bold to enable a simple reading flow. Specific bottlenecks along with information on the selected pathway are given. All active processing steps are summarized in Table C.9 of Appendix C.5. Table 5.2 presents detailed results of the extreme points.

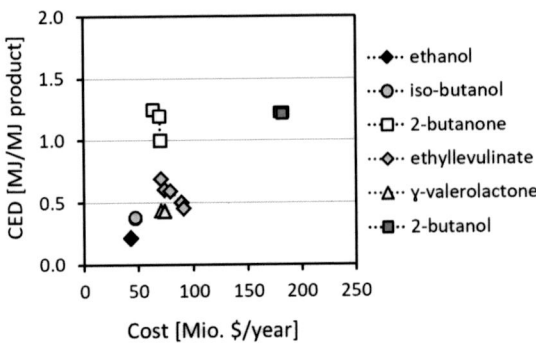

Figure 5.3: PNFA results (Eq. 3.65) with Pareto curves demonstrating a different process pathway or separation strategy at every point.

Besides ethanol, the production of **iso-butanol** is identified as promising. Since only one possible pathway exists, the Pareto curve is a single point. While ethanol is produced from cellulose and hemicellulose, iso-butanol is only obtained from the cellulose fraction. Hence, the production of iso-butanol can be fully competitive to ethanol in the future, if an efficient exploitation of the hemicellulose fraction is enabled. Since iso-butanol cannot be produced from hemicellulose, alternative products, which can be used in a fuel mixture with iso-butanol, need to be identified. Due to a missing hemicellulose exploitation, the performance differences between ethanol and iso-butanol mainly result from higher waste disposal cost for the production of iso-butanol. Although the separation of ethanol from water is more challenging compared to iso-butanol, VRC reduces the energy demand for the ethanol separation significantly. At the same time, VRC increases the IC, such that the IC for ethanol production are higher compared to iso-butanol.

The production of **2-butanone** is economically competitive, but exhibits a high energy demand. Similar to iso-butanol, only one possible pathway exists but several separation strategies are analyzed. The major bottleneck is an energy-intensive concentration increase of 2,3-butanediol in water. Further alternative separation strategies as proposed by Penner et al. (2017) are key for future improvements.

5.1 Single-product biorefinery for fuel production

The results for **γ-valerolactone** reveal a change from the RNFA pathway (R29, R40) to the PNFA pathway (R5, R33) at the point of minimum cost. The RNFA is only based on the reaction yield, which is 72% starting from glucose (R29, R40). Although the PNFA pathway (R5, R33) offers a pathway yield of only 50%, separation is simplified since an intermediate separation of levulinic acid, formic acid and water (R40) is skipped. In addition, the water content in the product mixture is reduced by nearly 40%. However, in reaction R33 γ-butyrolactone is used as reaction solvent, which exhibits similar properties like γ-valerolactone. Due to a minor difference in the boiling points only, VRC can be efficiently applied. Hence, additional investment costs for the intermediate separation as well as utility cost for separation enforce the pathway change from the RNFA to the PNFA. The distinct difference in economic performance between ethanol and γ-valerolactone is also interesting since the RNFA assesses both processes as competitive.

The Pareto front for **ethyllevulinate** shows an interesting phenomena, as a change of pathways occur. This leads to a Pareto curve instead of only a single point. The minimization of TAC causes a pathway selection via hydroxymethylfurfural (R4, R5, R7, R8, R26). This pathway is shorter and requires less reactions and separations compared to the pathway via levulinic acid (R4, R5, R7 - R10), which is selected when minimizing the energy demand. Therefore, it does make a difference if a cost optimal pathway or one with a low CED_{fuel} is chosen. Since ethyllevulinate production requires a stoichiometric amount of ethanol, the number of active steps and hence the investment cost are higher compared to all residual fuels.

The production of **2-butanol** is by far the most expensive process. Only minor deviations occur between the point of minimal TAC and minimal CED, such that a Pareto curve exists, but is hardly visible. Similar to the RNFA, the major reason for the costly process is an unselective final reaction to 2-butanol (R34). A pathway via R36 is even more expensive due to a very low yield. While a consideration as single-product process can be excluded, a performance improvement is obtained in case the side-products (2-methyltetrahydrofuran, 1-pentanol) of R34 are included in the final fuel as well.

In Table 5.2 further results including the GWP, the carbon (CE) and combustion efficiency (η_{comb}) are given. The non-renewable GWP follows the same pattern as the CED. Fossil gasoline exhibits a non-renewable GWP of 89 $\frac{gCO_{2,equivalents}}{MJ_{fuel}}$ (International Institute for Sustainability Analysis and Strategy, 2016), which renders the production processes of all products except 2-butanol promising. However, biomass cultivation, harvesting and the biomass supply chain are not included in these numbers yet.

5 Application of PNFA

Table 5.2: Detailed results of the extreme points on the Pareto curve for all products. For ethanol and iso-butanol the point of minimum TAC equals the point of minimum CED, such that only a single row is given. For the residual products, the first row refers to the minimization of TAC and the second row to the minimization of CED.

Product	min.	Cost $\frac{\text{Mio.\$}}{\text{year}}$	IC Mio.$	CED $\frac{\text{MJ}}{\text{MJ}_{fuel}}$	GWP $\frac{\text{gCO}_{2eq.}}{\text{MJ}_{fuel}}$	CE $\frac{\text{mol C}_{fuel}}{\text{mol C}_{Bio}}$	η_{comb} $\frac{\text{MJ}_{bio}}{\text{MJ}_{fuel}}$
ethanol	Cost	43	147	0.22	16	0.46	1.5
iso-butanol	Cost	46	97	0.38	25	0.32	2.3
2-butanone	Cost	64	133	1.25	81	0.38	2.1
	CED	69	150	1.00	74	0.38	2.1
ethyllevulinate	Cost	70	238	0.69	45	0.43	2.1
	CED	91	336	0.46	31	0.42	2.1
γ-valerolactone	Cost	70	156	0.43	32	0.29	3.1
	CED	74	176	0.43	32	0.29	3.1
2-butanol	Cost	178	145	1.23	91	0.08	8.6
	CED	181	161	1.21	91	0.08	8.6

The analysis of the carbon and combustion efficiency reveals interesting aspects. The carbon efficiency (CE) of the processes to ethyllevulinate (43%) and 2-butanone (38%) are nearly as high as the CE of ethanol (46%). Similarly, the combustion efficiency of 2.1 for 2-butanone and ethyllevulinate are closest to the combustion efficiency of ethanol (1.5). Nevertheless, the Pareto curves show a distinct cost and CED difference between these processes and ethanol. The reason for the ethyllevulinate process are the high number of steps, which cause high IC of 238 Mio. $. This illustrates that the process to ethyllevulinate is on the one hand very selective, but on the other hand very costly. Thus, a summary of multiple steps into a single reaction is most promising for a future improvement. For 2-butanone, alternative separation strategies which result in a lower utility cost are key.

Furthermore, the CED minimization does not lead to significant improvements neither in the carbon nor in the combustion efficiency since most pathways remain similar, while only the type of separation is altered. Compared to previous results (Ulonska et al., 2016a), the consideration of utility cost increases the carbon and combustion efficiency for the production of ethanol and ethyllevulinate due to the conversion of the hemicellulose fraction into ethanol.

The hemicellulose fraction is not exploited for the residual products due to additional IC and an energy-intensive separation of furfurylalcohol and water after R16.

5.1 Single-product biorefinery for fuel production

This renders a further conversion of the hemicellulosic fraction inefficient. Hence, an analysis of alternative separation strategies is mandatory to improve the performance of the pathways starting from hemicellulose. Furthermore, a consideration of products that have a higher value than fuels is necessary.

Both the RNFA and PNFA are based on the assumption of a conversion limitation as a best case scenario (cf. Section 2.4.2). The selected pathways are all direct pathways. Parallel pathways do not exist. Therefore, the assumption of a conversion limitation is justified. Since mass preservation applies, a switch to a selectivity limitation does not alter the results significantly.

Scenario II: PNFA results with internal energy supply In scenario II), the PNFA lowers the number of process fluxes from 116 down to 44 and the number of reaction fluxes from 33 to 14. Similar to scenario I), a significant reduction of the number of reactions is achieved compared to the RNFA.

With an internal energy supply, the combustion of residual biomass, lignin and hemicellulose results in an external energy requirement CED of zero for all biofuels, which equals the lower bound. Hence, the Pareto curves can no longer be represented as trade-offs between minimization of TAC and CED since all points on the Pareto curves coincide. A variation neither of the cost nor of the energy demand exists.

The only exception is the production of ethyllevulinate with a CED of 0.06 $\frac{MJ}{MJ_{fuel}}$ at the point of minimal cost. The external electricity demand of this process is low, which makes it economically unattractive to buy a turbine to offset the utility cost. Similar to all other processes, the CED is zero at the point of minimal CED.

Compared to scenario I), the pathways to iso-butanol, 2-butanone and γ-valerolactone remain the same. In case of ethanol and ethyllevulinate, the hemicellulose fraction is no longer converted into ethanol, but sent to combustion as well. More reactions are active for the production of 2-butanol, which leads to higher investment costs, but lower raw material and waste disposal cost. This change is enabled due to a significant reduction of the utility cost, which normally prevents the activation of pathways exhibiting several energy-intensive separations.

In order to analyze whether it is necessary to distinguish between the scenarios, Figure 5.4 presents a ranking of the biofuels based on the production cost per volume and energy content for both scenarios. The results are given for the point of minimal cost. Both scenarios reveal the same fuel ranking and similar specific cost for each fuel. The reason for the similar cost is a compensation of the additionally required investment cost (boiler, turbine) by the reduction of utility and waste disposal cost.

5 Application of PNFA

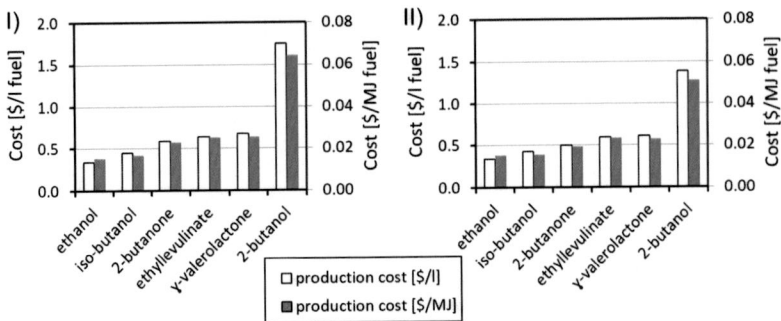

Figure 5.4: Fuel cost comparison for scenario I) excluding (left) and scenario II) including an internal energy supply (right).

An exception is 2-butanol for which a significant cost reduction is achieved in scenario II). However, this does not influence the overall statement, that 2-butanol is not competitive. Furthermore, the relative cost difference between 2-butanone on the one hand and ethyllevulinate and γ-valerolactone on the other hand is enlarged revealing a superior performance of 2-butanone. The analysis of alternative separation strategies for the 2-butanone pathway might favor this process in the future. This statement is strengthened if the volumetric energy densities are utilized as basis for comparison. The percentage cost difference between ethanol and 2-butanone is reduced from 75% to 49% (scenario I) and from 49% to 27% (scenario II) due to a comparably higher energy density of 2-butanone.

Besides similar production cost in both scenarios, obviously scenario I) allows a more detailed analysis of the different process opportunities. The key message of scenario II) is that the external energy demand can be reduced to zero if additional or excess biomass (residues) are burned. However, burning waste residues superimposes the different pathway opportunities such that only one pathway is chosen for all points on the Pareto curve. A clear differentiation between the pathways and, thus, an identification of the major influence factors is not possible. In addition, the resource exploitation is lower. Finally, a higher value chain contribution than grid energy is targeted for the biomass residues. Therefore, the following results are all based on scenario I) excluding biomass (residue) combustion.

5.1 Single-product biorefinery for fuel production

Product ranking comparison In a literature study by Tao et al. (2014) the processes for ethanol, iso-butanol and 1-butanol are compared and ranked. This study is used to compare the rank order obtained with the PNFA. For this purpose, the PNFA results of 1-butanol are used here, a detailed description of which is presented in Section 5.2. The PNFA shows that the process for 1-butanol is more energy intensive and costly compared to ethanol and iso-butanol. According to Tao et al. (2014) ethanol production outperforms iso-butanol and 1-butanol in terms of energy demand and minimum selling price (MSP) as well. Thus, the product ranking, both in terms of energy demand and production cost, is the same using the PNFA. Therefore, a major PNFA goal of ranking production performances is successfully achieved.

Table 5.3 presents a yield and cost comparison. The same daily feed stream of 2000 dry metric tons as given by Tao et al. (2014) is utilized here, whereas parameters like feedstock composition, prices or yields are not fully disclosed. In addition, it is not clear whether 1-butanol production considers the effect of acetone and ethanol co-production (cf. Section 5.2). Thus, the comparison gives a first impression but needs to be handled with care.

Table 5.3: Comparison of PNFA process data with literature data by Tao et al. (2014).

	units	ethanol		iso-butanol		1-butanol	
		literature	PNFA	literature	PNFA	literature	PNFA
Output	Mio. L/y	231	345	170	197	140	108
Yield	L/t_{feed}	330	493	242	281	200	154
MSP	$/L	0.57	0.28	0.78	0.40	0.80	0.72

The PNFA reveals a higher yield for ethanol and iso-butanol production and, thus, a lower MSP compared to the data by Tao et al. (2014). Besides raw material costs, the cost structure for iso-butanol production presented by Tao et al. (2014) reveals pretreatment and enzymes cost as major contributions. Therefore, the same aspects (pretreatment, enzyme cost) as determined in Chapter 4 explain the lower MSPs obtained with the PNFA here.

Cost structure The cost structure of the processes are discussed and compared to conventional and biorefinery processes from literature. Table 5.4 summarizes the cost composition for all products at the point of minimal TAC. Results are shown for both combustion scenarios for the sake of completeness. As described above, the TAC are similar as scenario I) requires more utility and waste disposal cost, while scenario II) exhibits higher IC due to the additional equipment for the boiler and turbine.

Table 5.4: Cost structure at the point of minimal total annualized cost.

	ethanol	iso-butanol	2-butanone	ethyllevulinate	γ-valerolactone	2-butanol
Scenario I						
TAC [$\frac{\text{Mio.\$}}{\text{year}}$]	42.6	46.3	63.5	70.1	70.5	178.5
Annualized IC	51%	31%	31%	51%	33%	12%
Feedstock	29%	39%	26%	24%	35%	39%
Auxiliary	-	-	-	-	-	10%
Utility	15%	18%	37%	19%	19%	21%
Waste disposal	4%	12%	6%	6%	13%	18%
Scenario II						
TAC [$\frac{\text{Mio.\$}}{\text{year}}$]	42.4	43.6	54.0	65.5	63.6	141.5
Annualized IC	58%	38%	51%	54%	46%	29%
Feedstock	33%	42%	39%	29%	39%	45%
Auxiliary	-	-	-	-	-	12%
Utility	4%	11%	8%	13%	4%	5%
Waste disposal	5%	9%	2%	4%	11%	9%

For all fuels and both scenarios, the annualized investment costs contribute between 31% and 58% to the TAC, while feedstock cost are the second major contribution accounting for 25%-45%. Except for 2-butanol, the auxiliary costs are zero, as no additional hydrogen is required for the residual pathways and a full recycle of solvents is achieved by sharp splits in the separations. In case of 2-butanol external hydrogen is stoichiometrically required for the hydrogenation of γ-valerolactone to 2-butanol. Depending on the type and efficiency of separation, the utility cost range between 15% and 37% in case internal energy supply is prohibited. Since the external utility requirement can be efficiently reduced by combustion, the utility cost are low with 4%-11% in scenario II). Similar to the utility cost, waste disposal cost are reduced in scenario II) as residues are (partially) burned.

In petrochemical industry, the cost structure of bulk chemicals normally deviates from the above described cost distribution. Cheali et al. (2015b) report a ratio of 80%-90% operating cost and only 10%-20% for the annualized investment cost. However, the cost structures obtained with the PNFA show a higher relative contribution of the annualized investment compared to the operating cost. Indeed, the investment cost correlation is a simplification which exhibits uncertainties. In order to analyze whether the high IC contribution stems from this correlation or is a general biorefinery problem, the cost distribution is compared to processes from literature. In Table 5.5 the cost distribution of processes for the lignocellulosic-based production of ethanol (Humbird et al., 2012) and ethyllevulinate (Hayes et al., 2006) are presented. Herein, the annualized IC and operating cost are given. The latter includes the feedstock cost, which is listed additionally.

Table 5.5: Cost structure of two biorefineries for lignocellulosic biomass and residues

	Unit	Humbird et al. (2012)	Hayes et al. (2006)
Feedstock		corn stover	paper sludge, agricultural residue
Product		ethanol	ethyllevulinate
Production output	$[\frac{t}{year}]$	$182 \cdot 10^6$	$133 \cdot 10^3$
Capital investment	[Mio.\$]	422.5	150
TAC	$[\frac{Mio.\$}{year}]$	128	66
Annualized IC	$[\frac{Mio.\$}{year}]$	63 (49%)	22 (34%)
Operating cost	$[\frac{Mio.\$}{year}]$	65 (51%)	44 (66%)
Feedstock cost	$[\frac{Mio.\$}{year}]$	45 (35%)	27 (41%)

Humbird et al. (2012) describe a detailed process design of a biochemical conversion of corn stover to ethanol. With a capacity of $182 \cdot 10^6 \frac{t}{year}$, the production output is orders of magnitudes higher compared to the PNFA results ($100 \cdot 10^3 \frac{t}{year}$).

5 Application of PNFA

Thus, an even larger benefit from the economy of scale is expected for the ethanol plant proposed by Humbird et al. (2012). However, as shown in Table 5.5, the IC still contributes to 49% of the TAC, while feedstock cost account for 35%. This is in line with the findings of the PNFA.

Similarly, the ethyllevulinate production process and the associated cost structure by Hayes et al. (2006) compare well to the results obtained with the PNFA though a slightly higher IC and lower feedstock contribution are determined in the PNFA. The capacities are in the same order of magnitude, such that the economy of scale does not influence the overall statement. Therefore, the shift in cost structure from an operating cost dominated production in case of petrochemical bulk chemicals, to a more or less equal contribution of IC and feedstock cost for biorefinery processes, is demonstrated here. This means that biorefinery processes have a higher capital risk and do not benefit from the economy of scale in the same way as petrochemical processes, since a capacity increase leads to higher biomass transportation costs. This has the consequence that biorefinery process development needs to focus on the reduction of IC, e.g., by establishing one-pot reactions instead of maximizing the reaction yield of a single conversion step.

5.1.5 Sensitivity analysis

As an early stage screening method, both the RNFA as well as the PNFA rely on parameters and assumptions, which are subject to uncertainty. Thus, the PNFA is integrated with a complementary sensitivity analysis (cf. Section 2.7) similar to previous studies (Ulonska et al., 2016b). For this purpose, the parameters are altered by ± 5% in a one-at-a-time analysis.

The sustainability parameters are comprised of the primary energy and global warming factors. The economic parameters include the IC parameters $Inv1$ and $Inv2$, the interest rate, run time, raw material, utility and waste disposal prices. The process parameters contain reaction yields, split factors for separations and the specific utility requirements of every separation. Reaction yields and separation split factors are bounded between zero and one. Since sharp splits for the separations are considered (cf. Section 3.2.1.2), the base value of the split factors equal one such that only a deterioration is feasible. In total, 372 parameters are altered in the sensitivity analysis.

A sensitivity analysis for a RNFA case study reveals $Inv2$ as most sensitive parameter (Ulonska et al., 2016b). This is explained by the nonlinearity of the exponential parameter. A distinction between the reaction performances has still been feasible as long as not all parameters are varied at the same time (Ulonska et al., 2016b).

5.1 Single-product biorefinery for fuel production

In the following, it is analyzed whether these statements can be transferred to the PNFA to clarify if the ranking of pathways is reliable for a given level of uncertainty and to identify the most influential parameters.

Figure 5.5 presents the sensitivity results at the point of minimal cost and minimal CED for the PNFA excluding combustion. A similar ranking and differentiation between the products compared to the base case (Figure 5.3) is obtained.

A clear differentiation of the worst performing 2-butanol and the residual products is possible even under the assumed uncertainties. This proves once more that 2-butanol production is not promising. The high CED deviation for 2-butanol at the point of minimal cost is remarkable, which results from a change of pathways replacing reaction R33 with reactions R8, R9 and R29. The reason for this change is a split factor reduction of the intermediate purification of γ-valerolactone from the reaction solvent γ-butyrolactone. While the alternative pathway is more cost efficient compared to the original pathway, it exhibits a higher energy demand due to multiple separations.

Compared to 2-butanol, the residual fuels exhibit only minor deviations under uncertainty. This can be explained by (i) a single outstanding pathway for ethanol, iso-butanol and 2-butanone and by (ii) several equally promising pathways for ethyllevulinate and γ-valerolactone.

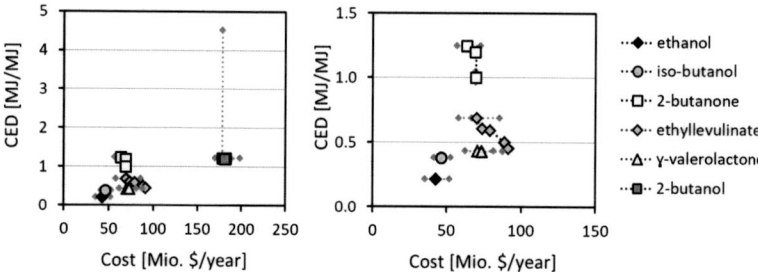

Figure 5.5: Results sensitivity analysis for all biofuels (left) and all biofuels except 2-butanol (right).

In Figure 5.6 the maximal TAC and CED deviation are presented for each fuel along with the parameter, which causes the largest deviations. The relative cost deviation is higher compared to the relative CED deviation. This is mainly caused by the nonlinearity of the IC parameter *Inv2*, which is again the most sensitive parameter.

5 Application of PNFA

An increase in the accuracy of this parameter is difficult, since the exponent in an IC function deviates from plant to plant. If processes are established, the capacity exponent can be transferred to similar plants. However, since novel processes are analyzed here, similar plants and a priori process knowledge is rarely available. Another reason for the high influence of the *Inv2* parameter is a relatively high IC contribution to the overall cost (cf. Section 5.1.4). If the *Inv2* parameter is excluded from the analysis, the most cost sensitive parameter is the γ-butyrolactone price as it strongly influences both the γ-valerolactone as well as the 2-butanol pathway.

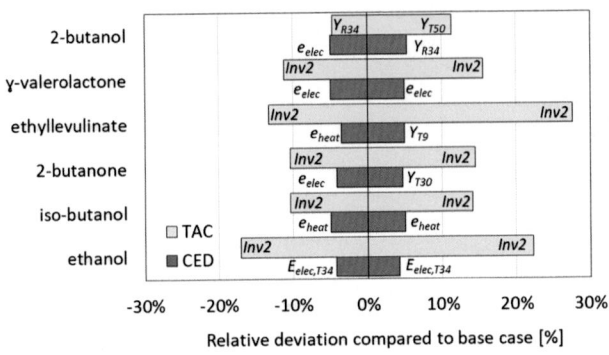

Figure 5.6: Relative deviation of TAC and CED compared to the base case. For every deviation, the most influential parameter is given as well.

In general, the deviation at the point of minimal CED is lower and a maximal deviation of 5% is observed. This is explained by the strictly linear functions for the CED determination. Besides the primary energy factor for heat and electricity, separation split factors or the energy demand for separations influence the results. In all cases the final purification steps are the most influential ones (cf. Appendix C.3 for the assignment of the separation numbers). The overall product ranking is not influenced by the considered uncertainties.

All of the previous results are obtained allowing a free pathway choice in every optimization run. If the pathways are fixed to the base case, similar results are obtained for most of the products. These are presented in Figure 5.7. Again a similar ranking and differentiation between the products is obtained.

Since the pathways can no longer be changed, the high CED deviation for 2-butanol observed in Figure 5.5 is significantly reduced. Remarkably high deviations are shown for γ-valerolactone and 2-butanol at the point of minimal cost. The reason in both

5.1 Single-product biorefinery for fuel production

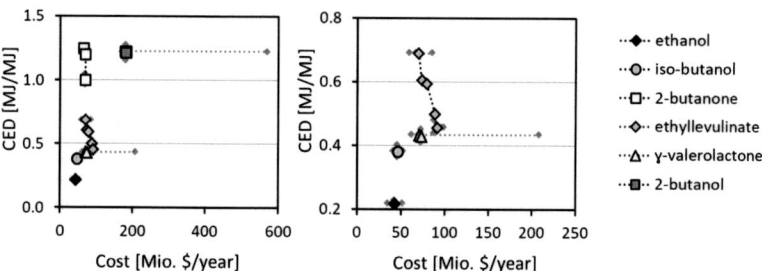

Figure 5.7: Sensitivity analysis for pathways fixed to the base value for all fuels (left) and all fuels except 2-butanol and 2-butanone (right).

cases is a reduced split factor for the separation of water from the mixture consisting of γ-valerolactone and γ-butyrolactone. If non-sharp splits are assumed in this distillation step, the mixture is partially diluted in the water stream and is lost. Therefore, the expensive solvent γ-butyrolactone needs to be added to the process, which increases the TAC. Thus, alternative reaction solvents need to be analyzed to reduce the high dependency of the process performance on the γ-butyrolactone price.

5.1.6 Heat integration

A simultaneous pathway selection and heat integration analysis increases the complexity of the optimization problem, which has then 684 equations, 736 continuous and 107 binary variables. This leads to an increase in the computational solving time compared to the single process optimization problem, e.g., for ethyllevulinate the computational time increases from 21 seconds to 38 seconds on an Intel(R) Core(TM) i5-6200 CPU (3.10 GHz). The results of the heat integration analysis are presented in Figure 5.8. It comprises the results of the PNFA, of a subsequent pathway selection and pinch analysis as well as the results of a simultaneous optimization.

Regardless of the optimization approach (subsequent, simultaneous approach), a significant heat integration potential cannot be observed for any of the fuels. Minor CED reductions are observed for the ethyllevulinate production and for 2-butanol using the simultaneous optimization approach. All results are verified manually and thereby validated. These hardly available CED improvements by heat integration are surprising since in petrochemical industry heat integration is the key for viable processes and significantly lowers the demand of external utilities (Biegler et al., 1997).

5 Application of PNFA

Therefore, the results are discussed in the following to identify the reasons.

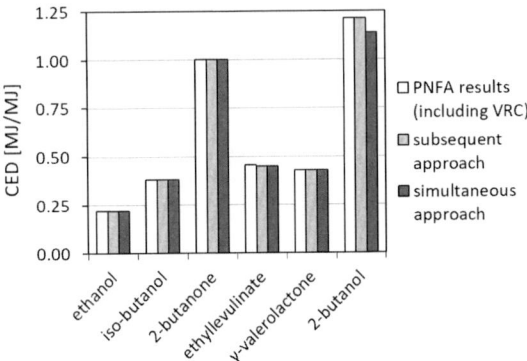

Figure 5.8: PNFA results for all products at the point of minimal CED along with pinch analysis results using a subsequent or a simultaneous approach of pathway selection and heat integration.

The first reason are short pathways. Since every separation costs energy, short pathways with as little as possible separations are preferred. This has the consequence that one-pot reactions, which combine multiple reaction mechanisms into a single step, are favoured. Furthermore, a single-product biorefinery is analyzed such that only a single pathway exists. Thus, only few heat sources and sinks are available.

The second reason is that energy-intensive separations are already selected using internal heat integration by means of VRC. If the utility cost for a standard separation is higher than the additional (annualized) investment cost of a compressor, a VRC separation has a cost advantage. When the CED is minimized, VRC separations are also preferred if the CED of the compressor is lower than the heat demand of the reboiler. Therefore, VRC separations are only selected for close-boiling mixtures like the separation of γ-valerolactone and γ-butyrolactone.

The third reason are similar temperature levels of the few remaining heat sources and sinks. Compared to the petrochemical industry, the temperature levels are lower and therefore the temperature differences are smaller in biorefineries. This is a general difficulty in biorefineries, which lowers the potential for heat integration.

A closer look into the production of ethyllevulinate illustrates the difficulties. The selected reactions to ethyllevulinate (R7-R10) are exothermic. The reactions R7 and R10 take place at ambient temperature, which bears no potential for heat integration.

5.1 Single-product biorefinery for fuel production

The reactions R8 and R9 take place at 238°C and 208°C. Two out of four separations are already selected as VRC separations. The reboiler (condenser) of the remaining two separations (ethyllevulinate - water - ethanol, levulinic acid - formic acid - water) operate at 206°C (80°C) and 257°C (100°C), respectively. Without a column pressure change, a condenser of one column cannot provide heat for the other column. Due to the large temperature differences between the condenser and reboiler, the economic viability and feasibility of such a pressure change is not given. Instead, reaction R8 can partially provide heat for the reboiler, which runs at 206°C. However, the enthalpy of reaction is substantially lower than the heat requirement of the reboiler. Therefore, only a minor CED reduction of 0.008 $\frac{MJ}{MJ}$ is achieved.

The CED reduction which is obtained for 2-butanol production results from the activation of an additional pathway via 2,3-butanediol to 2-butanol. While this leads to a slight increase in the TAC due to additional IC, a CED reduction of 0.075 $\frac{MJ}{MJ}$ is achieved. For the residual products, no pathway changes occur. Therefore, neither the results nor the product ranking or pathway selections are substantially influenced by the pinch analysis. Herein, the subsequent heat integration approach is sufficient, while an additional integration of a heat exchanger network into the optimization does not promise further knowledge gain due to minor achievable CED improvements.

The example illustrates the difficulties of heat integration for single-product plants. In order to enable further energy integration in biorefineries, multi-product plants are targeted. Multiple parallel pathways enhance the probability of heat integration possibilities since the number of heat sources and sinks are increased. In addition, heat pumps might improve the potential for heat integration, especially in case of small temperature differences between hot and cold process streams.

5.1.7 Biomass supply chain design

The influence of biomass transportation and a price-market dependency for biomass, are analyzed for potential biorefineries located in the region of North-Rhine Westphalia (NRW), Germany. The available biomass harvesting area needs to be identified, which is presented in Appendix C.6 for brevity reasons.

The discretization of the harvesting area influences the accuracy of the results and the computation time for solving the optimization problem. In order to determine a sufficient resolution that allows for appropriate computation times, a brief analysis of various biomass harvesting cell sizes is conducted.

5 Application of PNFA

In a second step, the optimization problem for an optimal biorefinery location and process network is conducted for a single-product biorefinery. This allows for a direct comparison with the previously obtained PNFA results and an evaluation of the biomass transportation influence on the economics and sustainability of the biorefinery process. For this purpose, the optimization problem 3.67 is solved.

Figure 5.9 shows the forest area of NRW along with ten chemical site locations, which are considered in this thesis. The selection is based on the decision criteria by Clausen et al. (2015) who propose to consider the distance to federal roads and freeways, the availability of full service offers, a harbour, a rail connection, a similar production field and research and development facilities as criteria for the selection.

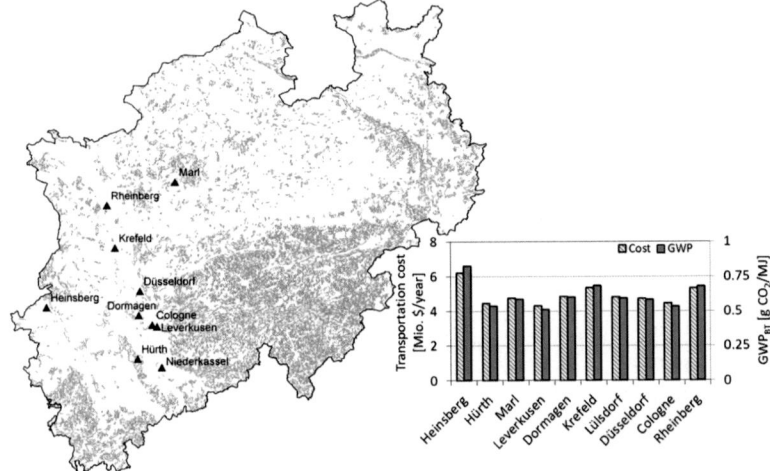

Figure 5.9: For the region of NRW, Germany, the distribution of available forest area (grey) as well as ten potential chemical site locations (▲) are shown. In addition, biomass transportation cost and GWP for an annual production of 100,000 t ethanol are determined for every plant location.

The results show that process data as well as costs other than transportation are equal for all sites. The analysis identifies the sites in Leverkusen, Cologne and Hürth as superior. The cost difference is only 3% such that all three are considered equally good. The results are reasonable as these sites are located close to the largest area of biomass availability. High transportation costs are seen for Heinsberg, which is expected, since it is located close to the border in an area with sparse biomass availability.

5.1 Single-product biorefinery for fuel production

The transportation cost contribute to 9-12% of the overall cost, which is low compared to 20-40% in literature examples (Angus-Hankin et al., 1995; Kumar et al., 2006; Ekşioğlu et al., 2009). The major reason for this discrepancy is a relatively small plant capacity with an annual production of 100,000 t ethanol compared to plants with an annual ethanol output up to 508,000 t (Ekşioğlu et al., 2009). Due to a higher capacity, more biomass is needed, which is transported over larger distances and, thus, result in a higher transportation cost.

Compared to a $GWP_{process}$ of 16.01 $\frac{g\,CO_2}{MJ\,ethanol}$, the GWP_{BT} with values ranging between 0.5 - 0.8 $\frac{g\,CO_2}{MJ\,ethanol}$ is negligible. The analysis proves the applicability of the extended PNFA model since (i) reasonable plant locations are identified and (ii) transportation cost are only slightly lower compared to literature ranges.

5.1.7.1 Influence of grid size on accuracy of results and computation time

Since the choice of grid size influences the accuracy of the results, the computation time, and solvability, an analysis of the sensitivity of the results to these factors is performed. Here, a comparison of three different cell sizes is presented. Since the data on biomass availability is taken from Dieter et al. (2001), who use a grid size of 4 times 4 km, the same grid size is considered here. This results in 2300 cells for NRW. In addition to a grid length of 4 km, a grid length of 10 km, as used by You and Wang (2011), and a grid length of 20 km are used. With a grid length of 10 (20) km, there are 400 (113) cells, respectively. This shows that the number of cells can be reduced by a factor of more than 20 compared to a grid length of 4 km. For an ethanol production plant, Table 5.6 summarizes the results on the biomass transportation cost (TBTC), the global warming potential for transportation (GWP_{BT}), as well as the average and maximal distance distance D for the three grid sizes. Furthermore, the required CPU times are listed.

Table 5.6: Influence of grid size on computational effort and results at the point of maximal revenues, while all points on the Pareto Curves lead to equal results.

	harvest sites	TBTC	GWP_{BT}	average distance	maximal distance	CPU time
	[-]	$[\frac{Mio.\,\$}{year}]$	$[\frac{gCO_{2eq.}}{MJ}]$	[km]	[km]	[s]
Without combustion						
4 km x 4 km	2300	4.62	0.489	43	63	2750
10 km x 10 km	400	4.62	0.489	44	65	124
20 km x 20 km	113	4.76	0.510	47	67	11

5 Application of PNFA

A decrease from a grid length of 4 to 20 km, leads to a CPU time reduction by a factor of 250, more than two orders of magnitude. This illustrates the importance of the grid size choice. A comparison of the results for a grid length of 4 and 10 km shows no differences for the TBTC and GWP. For the largest grid length of 20 km, only minor deviation of 3% in TBTC and 4% in GWP_{BT} are observed. Even the values for the maximal and average D differ by only 4 km. All these differences are well within the uncertainties of the underlying data and model assumptions.

In order to ensure that the method works, Figure 5.10 illustrates the resulting biomass supply area for the different cell sizes. The dark grey cells indicate the biomass harvesting areas. The white spots in Figure 5.10 a) indicate cell sizes without biomass availability. The optimal plant location is marked as black triangle.

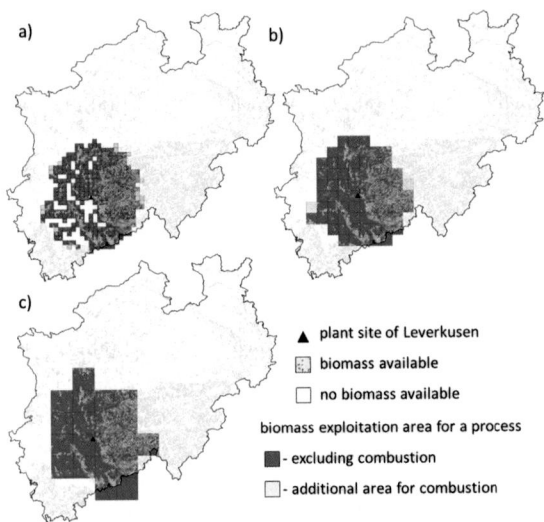

Figure 5.10: Available biomass potential (light grey) and supply radius for ethanol production in dependency of the edge length of a) 4km, b) 10 km and c) 20 km. The optimal site location (Leverkusen) is shown (triangle).

Leverkusen is identified as optimal plant location in all cases. The biomass is collected nearby. While for a cell length of 4 km, a more detailed presentation of the biomass harvesting area is possible, similar supply radius and biomass collection area are determined with larger cell sizes.

5.1 Single-product biorefinery for fuel production

Leverkusen is located close to the main biomass availability area and is therefore expected to be optimal. In addition, a similar supply radius is obtained for all grid lengths, which uses the available biomass close to the plant location. This confirms a proper working of the method.

For the sake of completeness, the results for scenario II) including an internal energy supply are given as well. The results show an increase in biomass supply and biomass exploitation of the cells in the edges of the resulting region. Neither the choice of plant location, nor a significant change of the supply radius is observed. Also no huge differences between the three grid lengths are obtained. Since only minor deviations between the cell length of 4 and 20 km are observed, a cell length of 20 km is deemed sufficiently accurate and computationally favorable and is consequently used for all further calculations.

5.1.7.2 PNFA results including supply chain

Once data and investigations on biomass availability, plant locations and grid length are available, the PNFA results including biomass supply chain are generated. For this purpose, the optimization problem 3.67 is solved. The optimization simultaneously analyzes the profitability and, thus, economic viability of the processes and supply chains.

Figure 5.11 illustrates the optimization results for all products and their processing pathways considering either a weak (a-b) or a strong (c-d) market scenario. This means that either a weak or strong market response is obtained with increasing biomass exploitation. The arrows in Figure 5.11 indicate the optimization goal of a low global warming potential (GWP) and less negative profit (TP), meaning that no single-product process achieves profitability independent of the market scenario.

For 2-butanol, 2-butanone, and ethyllevulinate, Pareto curves are shown in the weak market scenario (cf. Figure 5.11 a) and b)), while for the remaining three candidates these collapse into a single point. The reasons for this are diverse. The production of ethanol is the same for both objectives. For iso-butanol there exists only one feasible pathway (R5-R38). For γ-valerolactone the pathway via the reactions R5 and R33 outperforms all alternatives considering a weak market response (cf. Figure 5.11 a) and b)). Due to an increase in cost pressure, short pathways are preferred for ethyllevulinate and γ-valerolactone, which directly reduces the number of separation steps and hence the energy demand.

Compared to process optimization (cf. Section 5.1.4), TAC rise by 10% for ethanol to 38% for 2-butanol, when biomass supply and price development are analyzed.

103

5 Application of PNFA

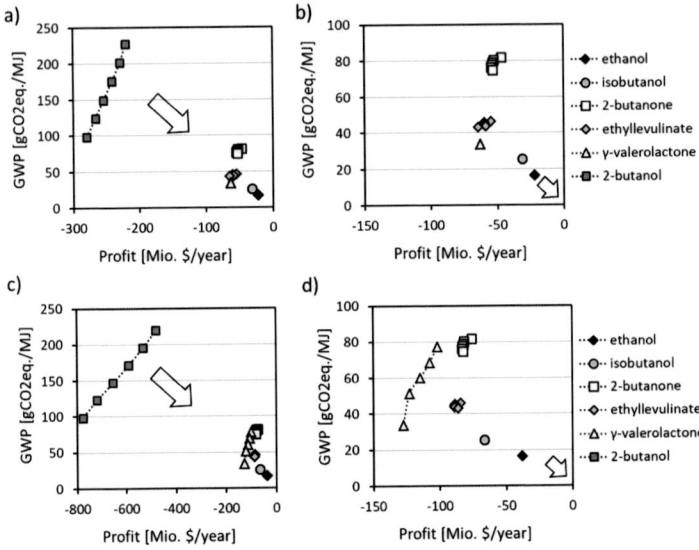

Figure 5.11: Pareto curves for a weak (a-b) and strong (c-d) market response on biomass market exploitation. While in the left part of the figure all results are shown, the right part presents a more detailed view for the top performing products of figures a) and c), respectively.

While the main reason is the high impact of transportation, roughly one third of the cost increase is caused by a higher biomass price. However, the effect of transportation on GWP is rather small with only 3% for ethanol to 7% for 2-butanol. Hence, process improvement is key for future GWP reductions.

The utilized fraction of available biomass varies strongly between 16% for ethanol and 91% for 2-butanol. While Leverkusen is identified optimal for all Pareto optimal solutions, different supply radii are required for the products. A maximal distance of 67 km is sufficient for ethanol, but for 2-butanol a maximal (average) distance of 127 (83) km is determined at the point of maximum profit and 193 (110) km at the point of minimum GWP. With the latter being the only exception, all other processes require distances of less than 150 km. According to Hamelinck et al. (2005), Mahmudi and Flynn (2006) and Kappler (2008), this confirms an economic transportation by truck only.

5.1 Single-product biorefinery for fuel production

The major difference between the weak and strong market scenario is the (negative) profitability of the processes, while the GWP as well as the pathways remain almost similar. The only exception is the production of γ-valerolactone. Here, an additional pathway is activated in the strong market scenario (cf. Figure 5.11 c) and d)). This is caused by a higher cost pressure, which requires a higher feedstock conversion.

All of the previous results are presented for scenario I) excluding biomass (residue) combustion. A similar ranking including (cf. Figure 5.12) and excluding combustion (cf. Figure 5.11) is determined for a weak biomass market response. While the combustion allows for a reduction in utility cost, these savings are partially compensated by an increase in raw material and transportation cost.

Furthermore, for carbon dioxide released during combustion, no waste disposal cost are determined. This leads to lower cost and hence a higher profitability for scenario II) including combustion. In addition, a lower GWP is observed in case of biomass combustion. This is caused by reduced external energy demand accompanied by the fact that carbon dioxide emissions caused from combustion are not considered in the non-renewable GWP.

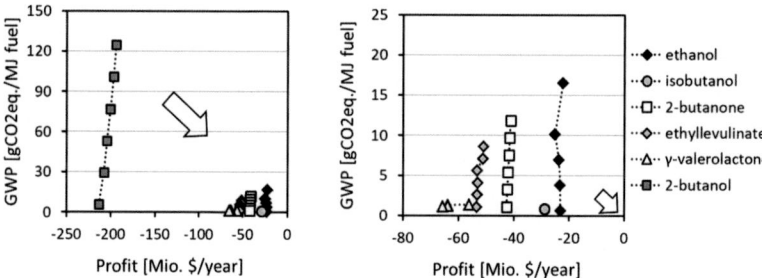

Figure 5.12: Single-product results for a weak market scenario allowing biomass and residue combustion for an internal energy supply. In the left part of the figure all products are shown, while the right side presents only the most promising results in more detail.

Although ethanol production requires less energy compared to the other processes at the point of maximal profitability, ethanol production exhibits a comparably high GWP. This is caused by a high resource exploitation such that less waste is available for combustion. At the same time, an increased biomass supply for combustion would lead to higher raw material cost, which are not compensated by reduced utility costs.

5 Application of PNFA

This example illustrates that the results considering combustion are more complicated to understand, while the main conclusions can be drawn from the results excluding combustion as well. Nevertheless, the combustion scenario demonstrates a non-renewable GWP reduction potential, but needs to be handled with care as it leads to significant carbon losses.

5.1.7.3 Sensitivity analysis

In order to evaluate how sensitive the obtained results and especially the process ranking are to uncertainty in the considered parameters, a sensitivity analysis is performed in combination with the optimization model. A parameter variation of ± 5% is applied for the following parameters: the transportation factors $P_{BT,fix}$, $P_{BT,var}$ and w_{H2O}, the initial biomass price $P_{initial,biomass}$, the utility prices for cooling water, steam and electricity $P_{utility}$, the waste disposal P_w and fuel selling price P_{fuel}, empirical investment costs parameters $Inv1$, $Inv2$, interest rate ir and plant run time n as well as the GWP factors for heat, electricity and transportation. Figure 5.13 illustrates the results for scenario I) excluding combustion.

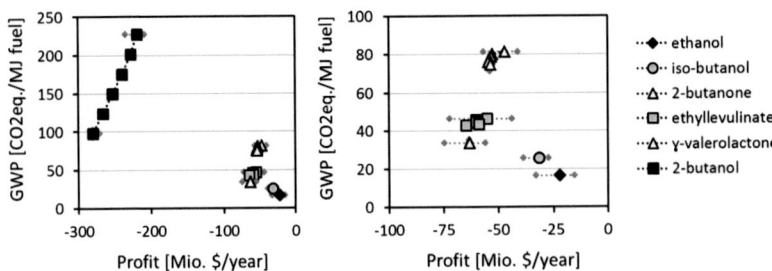

Figure 5.13: Results sensitivity analysis for all biofuels (left) and all biofuels except 2-butanol (right).

Only minor GWP deviations are observed, which are not even visible in Figure 5.13. Higher discrepancies occur for the profit and a partial overlapping. The (exponential) deviations in case of profit maximization are mainly caused by the variation of the parameter $Inv2$, which is in line with previous findings (cf. Ulonska et al. (2016b), Section 5.1.5). Next to $Inv2$, the parameters $Inv1$, P_{fuel} and $P_{initial,biomass}$ are the most influential parameters, which lead to a linear deviation of maximum ± 6%. The most sensitive parameters for GWP are the pre-factors gwp_{elec} and gwp_{heat}.

Since the influence of the energy demand on the GWP is reduced in a scenario considering residue combustion for an internal energy supply, the most sensitive parameter in that case is gwp$_{truck}$. However, only minor deviation of ± 6% occur due to a linear dependency of all three parameters for GWP. Most importantly, the ranking is not affected and the selected pathway remain valid.

5.2 Analysis of fuel mixtures

While in previous sections only pure products are discussed, fuel mixtures might be advantageous from a product as well as from a process perspective. From a product perspective, tailored mixtures can be advantageous in order to overcome cold-start problems, reduce emissions, improve knocking resistance or increase the fuel's heating value (Thewes et al., 2012; Storch et al., 2015; Dahmen and Marquardt, 2016).

Two approaches to obtain fuel mixtures are conceivable. A mixture can be produced by blending different pure components, which have been produced by diverse pathways or at least different final reactions. Compared to pure products and single pathways, a higher number of active processing steps is required, which increases the IC as well as the utility cost. Mutual benefit might occur, for instance, from heat integration between the pathways leading to an overall CED reduction. In the case study of Section 5.1, no fuel mixture is favorable, as no product performs better in any of the objective functions than the two most promising processes towards ethanol or isobutanol. A mixture might be enforced by constraining the mixture composition or setting a specific upper product limit to fulfill certain product specifications.

A second opportunity to produce mixtures are unselective reactions, which automatically generate a product spectrum that can be directly used. A final product purification is skipped, which reduces the CED and separation cost. When the full product spectrum is considered, a more efficient production with a higher resource consumption is obtained compared to an energy-intensive purification of the mixture.

The following case study demonstrates the PNFA applicability for fuel mixtures. A mixture sufficient for spark-ignition engines consisting of acetone, 1-butanol and ethanol (ABE) obtained via hexose fermentation is analyzed. In addition, a mixture for compression-ignition engines (C8-C16 mixture) is presented. The latter is mainly composed of 1-octanol (OL), dioctylether (DOE) and butyltetrahydrofuran (BTHF) produced by aldol-condensation and further hydrogenation of furfural and acetone (Julis and Leitner, 2012; Luska et al., 2014). Acetone can be supplied internally by ABE fermentation or externally. Furfural is obtained from pentoses.

5 Application of PNFA

Two co-production scenarios are analyzed, describing a co-production of both mixtures as well as a subsequent production of di-n-butylether (DNBE). Together with the C8-C16 mixture, DNBE shows promising mixture behaviour in a compression-ignition engine (Heuser et al., 2014; Kerschgens et al., 2016).

Figure 5.14 presents a reaction network comprising all different pathways. The production challenges are briefly described before the process performances are discussed. Subsequently, both co-production scenarios are analyzed. Due to simplicity reasons, all of the following results do not consider an internal energy supply by waste combustion.

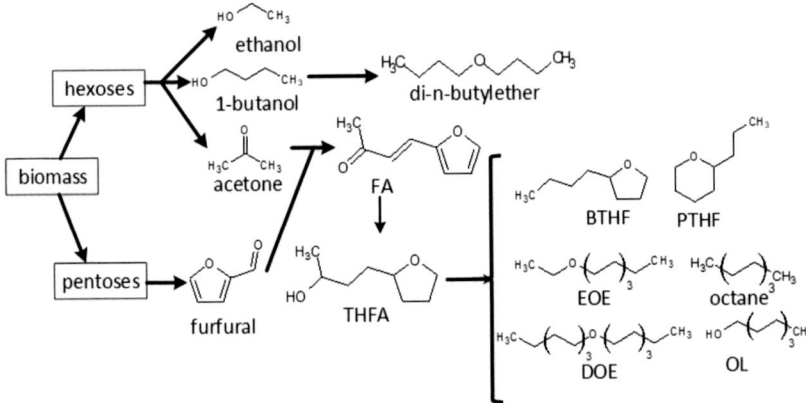

Figure 5.14: Reaction network for the production of mixtures.

ABE fermentation Even though the ABE fermentation is well known in literature and industry, its process still faces severe challenges. Known bottlenecks are an upper solvent concentration of around 20 $\frac{g}{L}$ leading to a highly diluted product. Difficulties in the solvent separation arise due to multiple solvent azeotropes with water (Green, 2011; Abdehagh et al., 2014). Since only a minor ethanol fraction is obtained, a challenging azeotropic ethanol water separation is skipped herein, thus reducing the energy demand as well as the IC. For the separation of 1-butanol, the miscibility gap with water is exploited, thereby lowering the effort in the interconnected distillation.

In general, a molar solvent composition of 3:6:1 (acetone:butanol:ethanol) is obtained (Green, 2011; Abdehagh et al., 2014). To enhance the 1-butanol fraction, a fermentation setup with a molar composition of 5:14:1 is utilized (Ezeji et al., 2007).

5.2 Analysis of fuel mixtures

A water content of 0.13 vol% (0.16 wt%, 0.6 mol%) is tolerated in the final ABE fuel, which is well below a reported upper tolerable limit of 0.5-1 vol% described by Chang et al. (2013) and Li et al. (2016). The water content is caused by skipping the expensive purification of the ethanol water azeotrope. Figure 5.15 presents the 1-butanol as well as the mixture results compared to a classical ethanol fermentation (cf. Section 5.1). Clearly, the utilization of the whole solvent mixture is favorable. The energy demand is reduced by 27% and the annual cost by 20% using the full ABE mixture instead of 1-butanol only. Even though, this demonstrates the improvement potential for mixtures compared to pure substances, the benchmark to a classical ethanol fermentation demonstrates that the process is still too costly and requires too much energy.

DNBE production Before analyzing a co-production of the C8-C16 mixture and DNBE, the DNBE process and associated PNFA results are briefly described. DNBE is obtained from 1-butanol as a starting reagent. No solvent is required in this reaction, which forms a stoichiometric amount of water. The subsequent separation of DNBE and water is simple, as they have a large miscibility gap. Hence, phase separation by decantation is possible. No DNBE is lost in the water phase, while only 1 wt.% water remains in the DNBE phase, which is assumed to have no significant influence on the fuel's engine performance. As shown in Figure 5.15, the production of DNBE is not competitive mainly due to the aforementioned bottlenecks in the ABE fermentation. A reaction yield of 92% (Khusnutdinov et al., 2012) leads to higher raw material cost and a slight increase in the specific CED compared to 1-butanol. Furthermore, the additional steps cause a rise of the IC.

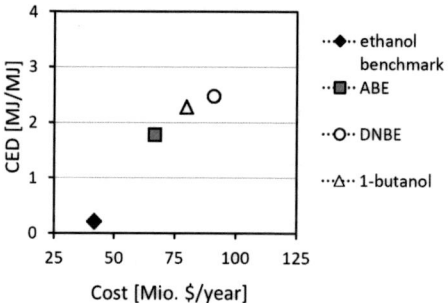

Figure 5.15: PNFA results for the ABE mixture benchmarked with the pure component fuels as well as the subsequent DNBE production.

5 Application of PNFA

C8-C16 mixture For the production of a mixture sufficient for compression-ignition engines, furfural and acetone undergo an aldol-condensation to form furfuralacetone (FA). This step is conducted in a biphasic setup, which consists of a water and a tetrahydrofuran phase (West et al., 2008). Acetone can be either supplied externally for a price of 0.81 \$/kg (Kumar et al., 2012) or produced internally by ABE fermentation. A surplus of acetone leads to an increased production of the single condensation product FA in comparison with the double condensation product.

FA is completely dissolved in the organic phase, while acetone reveals a partition coefficient of four, hence showing a higher solubility in the organic phase compared to the water phase (West et al., 2008). Based on the approach of thermodynamic insights (Jaksland et al., 1995), a product separation by evaporation is feasible as no azeotrope exists and the boiling point and vapor pressure ratios exceeds the lower limits.

In a subsequent reaction, FA is hydrogenated first to 4-(2-tetrahydrofuryl)-2-butanol (THFA) and then to a mixture consisting of OL, DOE, BTHF, propyltetrahydropyran (PTHF), ethyloctylether (EOE) and octane (Julis and Leitner, 2012; Luska et al., 2014). An intermediate separation between the second and third reaction is not necessary as no solvent is required and no by-products are formed. The final fuel mixture is separated from the solvent (ionic liquid) by extraction and subsequent evaporation of pentane (Julis and Leitner, 2012; Luska et al., 2014).

By changing reaction conditions of the third reaction, like the choice and composition of catalyst, reaction temperature and time, different product compositions are producible (Julis and Leitner, 2012; Luska et al., 2014). These product compositions along with their respective references are summarized in Table 5.7 for two alternative starting reagents (FA and THFA).

Table 5.7: Molar mixture compositions

#	substrate	x_{BTHF}	x_{PTHF}	x_{OL}	x_{DOE}	x_{EOE}	x_{Octane}	Reference
1	FA	0.05	-	0.50	0.45	-	-	(Julis and Leitner, 2012)
2	THFA	0.09	-	0.26	0.55	-	0.09	(Julis and Leitner, 2012)
3	THFA	0.03	0.03	0.42	0.46	0.06	-	(Luska et al., 2014)
4	THFA	-	0.02	0.32	0.62	0.03	0.01	(Luska et al., 2014)
5	THFA	-	0.01	0.25	0.66	0.07	0.01	(Luska et al., 2014)
6	THFA	0.82	0.01	0.13	0.04	-	-	(Luska et al., 2014)
7	THFA	0.14	0.01	0.60	0.22	0.03	-	(Luska et al., 2014)

5.2 Analysis of fuel mixtures

While in a first attempt Julis and Leitner (2012) used 38 $\frac{L}{\text{mole THFA}}$ of pentane (setup I), Luska et al. (2014) have been able to reduce this to 15 $\frac{L}{\text{mole THFA}}$ (setup II). First successful laboratory experiments even show an efficient extraction using 1.3 $\frac{L}{\text{mole THFA}}$ (setup III). As the fuel mixture consists of non-polar molecules, a second phase occurs after the reaction and therefore even a decantation might be feasible in the future (Luska, 2016). All four purification scenarios are analyzed with the latter representing a best-case scenario, which has not been demonstrated in laboratory yet.

Figure 5.16 presents a comparison of the final purification scenarios depending on the amount of pentane used for extraction. The purification scenarios have been analyzed for the same mixture composition (Table 5.7 entry 5). While in first experiments an excess of pentane has been used, the solvent reduction leads to a CED decrease of 40% in setup III compared to setup I. The decantation shows no significant improvement compared to setup III since the separation of the light boiling pentane (36 °C) from the high boiling product mixture is easily conducted using distillation. As the assumed sharp split between the ionic liquid and product mixture by decantation is not yet proven, the following results consider setup III only.

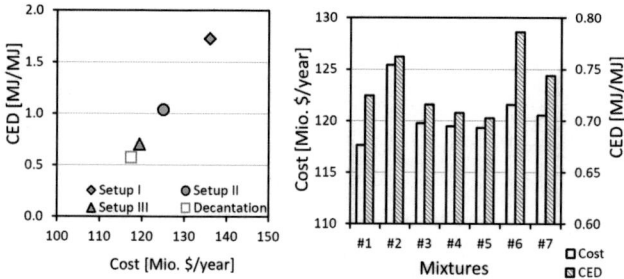

Figure 5.16: Left: Analysis of the final separation setup on the process performance. Right: Analysis of mixture composition on the process performance.

In the following, the influence of the mixture composition on the overall process performance is analyzed. For this purpose, seven different compositions (cf. Table 5.7) are compared. The Pareto curves are represented by a single point only, as for a single mixture no alternatives exist. Therefore, the mixtures are benchmarked based on their production cost and CED in a bar chart in Figure 5.16.

Mixture 1 constitutes the cheapest and mixture 5 the composition with the lowest CED. An annual cost difference up to 9 Mio. $\frac{\$}{\text{year}}$ and a CED difference up to 0.09 $\frac{\text{MJ}}{\text{MJ}}$ is caused by the different compositions. Hence, a thorough mixture analysis is important.

111

5 Application of PNFA

The main reason for the cost difference is a higher yield based on the substrate FA of 98% (mixture 1) compared to 92% (mixture 2). In addition, mixture 1 is directly obtained from FA while for the residual compositions a required intermediate step via THFA leads to higher IC.

The reason for the CED difference is the ratio of aliphatic ethers (DOE, EOE) to cyclic ethers (BTHF, PTHF) and octanol. DOE has a higher heat of combustion compared to the C8 fraction resulting from a higher number of carbon atoms and a higher hydrogen demand for hydration. Therefore, to fulfill the design constraint α, less THFA and consequently less furfural is required, which reduces the effort for the energy intensive separation of furfural and water. Hence, a lower CED can only be achieved by accepting a higher hydrogen demand.

In general, for all mixtures an external acetone supply is preferred over an internal supply by ABE fermentation as the latter increases the IC and utility cost as well as the CED.

In Figure 5.17, the C8-C16 mixture is benchmarked against their corresponding pure products and ethanol fermentation. For this reason, the mixtures exhibiting the highest amount of 1-octanol (mixture 7), BTHF (mixture 6) and DOE (mixture 5) are compared with the production of their pure components. An energy reduction of 17% (DOE) to 54% (1-octanol) can be achieved considering mixtures instead of pure products as the effort for an additional separation can be skipped. Compared to ethanol, none of the mixtures is competitive.

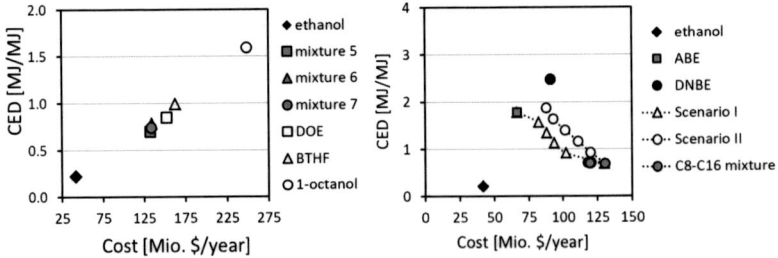

Figure 5.17: Left: Benchmark of pure component fuels with their corresponding mixture performance and ethanol from fermentation. Right: Co-production allowing a simultaneous production of ABE and a C8-C16 mixture (scenario I) or of DNBE and a C8-C16 mixture (scenario II).

5.2 Analysis of fuel mixtures

Main CED bottleneck is the purification of furfural. In the conversion of xylose to furfural the xylose concentration is kept low to prevent undesired side reactions (Marcotullio and de Jong, 2010). Even though the reaction solvent water does not need to be separated completely and a miscibility gap is formed by furfural and water, the concentration increase of furfural in water is energy intensive as water is the light boiling component.

Main cost bottlenecks are a production from pentoses only, a stoichiometric requirement of acetone and a hydrogen demand of five to nine mole hydrogen per mole product. Since acetone is produced in the ABE fermentation, an overall performance improvement might be obtained by considering the ABE mixture for spark-ignition engine and the C8-C16 mixtures for compression-ignition engine. The acetone obtained in the ABE fermentation then either remains in the gasoline mixture or is used as co-reactand for the aldol-condensation. This is analyzed in the following section.

Co-production scenarios Two different scenarios are described to analyze a potential benefit of co-production. In a first scenario, the hexose fraction is converted in the ABE fermentation and the pentose fraction in the aldol-condensation yielding the C8-C16 mixture. In a second scenario, 1-butanol further reacts to DNBE. The results for both scenarios are shown in Figure 5.17 and the product compositions are given in Appendix C.8.

In scenario I, the point of minimal cost equals a single ABE fermentation, while the point of minimal CED is similar to a C8-C16 mixture production. Therefore, no benefit from co-production is gained. In between these extrema, both process strains are active. In scenario II, the point of minimal CED equals again the C8-C16 mixture production. A significant CED reduction of 25% can be achieved at the point of minimal cost by a parallel production of DNBE and the C8-C16 mixture. Reasons are an internal acetone supply for FA production as well as a higher output compared to the production of DNBE only.

A pinch analysis affects the results only marginally as only few separation steps are active. Additionally, VRC variants of energy-intensive separations (here furfural water separation), are already chosen, which further lowers the number of available heat sources and sinks.

While the first scenario shows that co-production, converting both sugar fractions, does not necessarily lead to significantly improved process performance, the second scenario demonstrates large energy saving potentials. Nevertheless, an ethanol competitive production is not enabled and thus, an economic production not possible.

5.3 Multi-product biorefinery for a co-production of fuel and chemicals

In order to increase profitability of the processes and design viable processes, value-added chemicals are key. Here, an economically efficient biorefinery is targeted where ethanol is produced as fuel. Ethanol is chosen since it constitutes the best performing product and is established in today's fuel market. In the following, the most promising co-products available in the network of Figure 5.1 are identified under consideration of biomass supply chain management and market response.

Any novel biorefinery drastically impacts the local biomass market. While the fuel output is small compared to the overall fuel market, the production of value-added chemicals strongly alters these low volume markets. Being conservative, only the strong market scenario is pursued in the following. Furthermore, the systematic analysis allows for a break even analysis to define a minimum quantity of required value-added co-products.

In Figure 5.18 a) the Pareto curves for a co-production of fuel, i.e., ethanol and value-added chemicals are shown. At the point of minimal GWP, only ethanol is produced. These results are equal to those of the single product analysis (cf. Section 5.1.7.2). Since ethanol production has been identified as the process exhibiting the lowest GWP, any additional active step or pathway adds GWP emissions.

Nevertheless, a co-production of value-added chemicals enables higher (positive) profitability. A co-production of iso-butanol with a mass ratio of 1.9:1 (ethanol:iso-butanol) is required to break even. If 2,3-butanediol and furfurylalcohol are produced in addition, profitability is maximized. For this point, the cost structure, mass and revenue distribution are shown in Figure 5.18 as well. The cost structure exhibits a high fraction of raw material and annualized investment cost. Compared to conventional crude oil processes, a higher investment cost contribution is determined. This results from multiple parallel processing steps. Furthermore, the economy of scale effect is lower than in conventional industry, as the positive scaling effect is offset by significantly increasing raw material and transportation cost. Thus, only 38% of the available biomass is utilized. A further biomass exploitation is not economically attractive. In addition to higher costs, the chemical prices decrease due to a higher market occupancy.

In order to fulfill the design constraint α, ethanol is produced. Even though 45% of the mass output is associated with ethanol, it has a low contribution of 5% to the revenues, which is caused by a low P_{fuel}.

5.3 Multi-product biorefinery for a co-production of fuel and chemicals

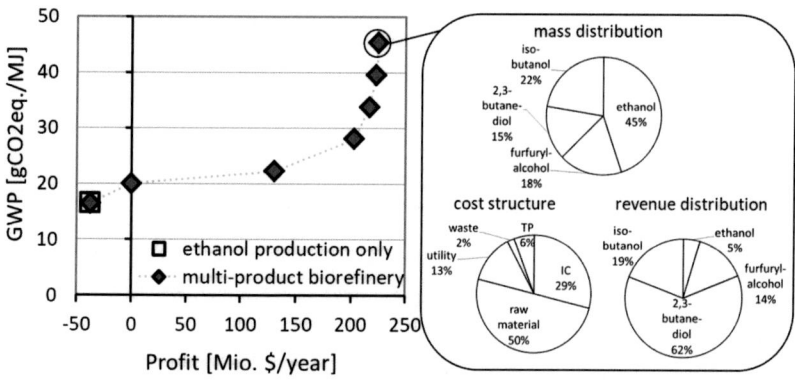

Figure 5.18: Multi-product biorefinery results for the co-production of ethanol (fuel) and chemicals assuming a strong market response. In addition, the cost structure, product mass and revenue distribution at the point of maximal profitability are presented. For comparison reasons, single-product results for ethanol are given as well.

The residual glucose fraction is converted into iso-butanol and 2,3-butanediol. The latter is responsible for 62% of the revenues, with only a share of 15% of the total mass output. However, a further 2,3-butanediol production would lead to a market oversaturation and therefore drastic price decline. Even though it increases the investment cost, this is the reason for a parallel iso-butanol production. The significantly larger iso-butanol market is also responsible for a more stable price.

Since the ethanol design constraint is achieved using the cellulose fraction, the hemicellulose fraction can be fully utilized for a value-added production of furfurylalcohol. In addition, the multi-product biorefinery benefits from the same logistics and biomass pretreatment for all products.

Overall, the results systematically prove a positive profitability of multi-product biorefineries co-producing fuel and chemicals based on a full model of biomass supply chain, process network and market analysis. However, main cost drivers like enzyme costs, wages and taxes are not considered, such that profitability will be reduced taking these factors into account.

115

As outlined by Kim and Dale (2003); Luo et al. (2009); Kaufman et al. (2010) and Ahlgren et al. (2015), GWP allocation is a major issue to determine the individual GWP contributions in a multi-product biorefinery. Herein, the partitioning method is applied, partitioning the GWP based on the mass, energy and exergy content.

For the point of maximal profitability, only small deviations are observed between the methods. The specific shares derived based on the enthalpy of combustion and on the exergy are similar, while only minor deviations for mass partitioning are observed. Ethanol has the largest GWP contribution with 44-45% of the total GWP, followed by iso-butanol (22-27%), furfurylalcohol (16-18%) and 2,3-butanediol (14-15%). Reason for a similar GWP partitioning independent of the partitioning method, is a similar ratio of roughly two (fuel:product) of the products molar mass, enthalpy of combustion and exergy. For this case study, the weighting choice does not influence the final results, but for multi-product plants with diverse products, the analysis is mandatory.

5.4 Summary and conclusions

The chapter demonstrates the applicability of PNFA for complex case studies. A fast evaluation of novel processes is accomplished using the PNFA. For a single-product biorefinery a comparison to RNFA results is given. Discrepancies between the RNFA and PNFA results are observed, which result from a change in the sustainability objective function and from the introduction of separations in the PNFA. Besides a product ranking, the process cost structure, the influence of residue combustion and heat integration are described along with a sensitivity analysis.

The PNFA of gasoline biofuels demonstrates that ethanol production is still superior compared to the residual analyzed alternatives such that no novel promising fuel production process can be identified. Beyond ethanol, the presented case study for gasoline biofuels reveals iso-butanol and 2-butanone as promising alternatives from a cost perspective. Due to non-selective final reactions, the 2-butanol production exhibits a high energy demand and can be excluded from further consideration as fuel.

The process ranking of ethanol, iso-butanol and 1-butanol are confirmed by comparison with a literature study of Tao et al. (2014). A cost structure comparison of the processes with literature known biorefinery processes confirms the obtained cost structure, with a relative high investment cost contribution compared to conventional processes.

5.4 Summary and conclusions

The CED reduction determined in the heat integration is lower than expected. The results are similar for the simultaneous and subsequent approach and are confirmed manually, which validates the method. Three major reasons are identified, namely (i) few process steps and therefore few available heat sources and sinks, (ii) heat-integrated distillations by means of VRC for energy-intensive separations and (iii) similar temperature levels for the heat sources and sinks.

The integration of a biomass supply chain identifies reasonable plant locations for the region of NRW, Germany. However, the analysis shows as well that economically attractive processes are not obtained.

Potential solutions are fuel mixtures and co-production scenarios. Even though a significant energy reduction by a co-production of DNBE and a C8-C16 mixture is obtained, a competitive production with ethanol is not achieved. In order to increase profitability, a co-production of value-added chemicals is examined. For such a multi-product biorefinery co-producing fuel and chemicals, an optimal product portfolio is determined stressing the importance of value-added chemicals to obtain profitability.

Overall, the results illustrate the potential of the PNFA methodology from biomass transportation via process analysis to product-portfolio selection.

Chapter 6

Biorefinery improvement potential

After demonstrating that the PNFA is able to provide reliable results in Chapter 4 and illustrating the application of the evaluation framework to single- and multi-product plants in Chapter 5, this chapter provides a more thorough discussion of the obtained results in order to derive (general) key improvement factors for the development of future biorefineries. Subsequent to the analysis of these key improvement factors in Section 6.1, the potential of fermentation processes is discussed in Section 6.2 as these addresses multiple of the developed key improvement factors and are of special importance in the transition to a bio-based economy.

6.1 Key improvement factors

The results discussed in the previous chapters, are used to derive general key improvement factors, which need to be addressed for an efficient development of future biorefineries. They are ordered in the sequence of the presented results, starting with the analysis of single- and followed by multi-product plants. At the end, additional aspects are discussed. Multiple aspects are derived from the single-product analysis, which are ordered according to the expected improvement potential starting with the largest impact. The criteria are similarly applicable for multi-product plants.

All criteria as well as the relevant sections of the PNFA results are listed in Table 6.1. Parts of the criteria have been known before or are general applicable for process design. Therefore, the importance in the context of biorefineries is stressed by discussing the results obtained with the PNFA.

6 Biorefinery improvement potential

Table 6.1: List of key improvement factors for the development of efficient future biorefineries derived from the results obtained with the PNFA.

Key improvement factors	Section(s)
valorization of both sugar fractions	5.1.4, 5.2
efficient solvent in particular water removal	5.1.3, 5.1.4, 5.2
reduction of investment costs	5.1.4
one-pot reactions	5.1.4
simultaneous capacity and supply chain optimization	5.1.7.2
energy integration techniques	5.1.6, 5.2
co-production of chemicals	5.3
additional resources and feedstocks	6.1
region-specific biorefinery solutions	6.1

Even though lignocellulosic biomass constitutes of lignin, hemicellulose and cellulose, often only the cellulose fraction is valorized. Due to complex molecular structures of lignin, valorization thereof is still an open issue such that it is used for an internal energy supply by combustion. However, a **valorization of both sugar fractions** is of high importance since the pentose fraction accounts for 20-40 wt% of the biomass (Huber et al., 2006). While the processes to ethanol and ethyllevulinate use both sugar fractions, the residual processes in Section 5.1.4 start from the hexose fraction, only. In addition, in the ethyllevulinate process the pentose fraction is converted into ethanol, which is required as co-reactand. A conversion via the pentose fraction and furfural is not selected for any of the processes in Section 5.1.4. Reason is an energy-intensive separation of furfural and water.

The products are often highly diluted in the solvents, exhibit high boiling points and form azeotropes. Thus, an **efficient solvent in particular water removal** is mandatory for viable processes. For this purpose, solvent changes along a pathway need to be prevented. Exceptions are conceivable, if the reactions take place in organic solvents as opposed to water, which is the starting solvent in hydrolysis. In these cases, water needs to be removed in particular if the reactions produce water as coproduct. However, thermal separations are often energy-intensive due to the high boiling products. A solution are bi-phasic setups, where the reactand is present in the water and the product in the organic phase. An ideal solvent exhibits a low water but high product solubility and has a boiling point lower than the product. In case of biotechnological conversions, the water removal is postponed to the end of a pathway. Since no additional solvent is required and intermediate separations are prevented, biotechnological conversions are promising and are further discussed in Section 6.2.

6.1 Key improvement factors

The analysis of the cost distribution of the optimized production processes (Section 5.1.4) indicates a shift from an operating cost dominated production in case of petrochemical bulk chemicals, to more or less evenly distributed shares of annualized investment and feedstock cost for biorefineries. Since a short payout time is required to convince potential investors to take the risk of increased investments in biorefinery processes, a **reduction of the investment cost** is necessary.

One attempt to reduce the IC are **one-pot reactions**, which combine multiple reactions into a single unit and thereby circumvent intermediate separations. Even though lower yields are obtained, the number of process steps and thereby the investment cost and energy demand are reduced. The example of the ethyllevulinate process in Section 5.1.4 illustrates the challenges which can arise from multiple reaction steps. While fermentation to ethanol requires a single reaction step, the ethyllevulinate process requires several reaction steps with intermediate separation. Although a similar carbon efficiency as for ethanol production is obtained for the ethyllevulinate process, the increased number of process steps, mandates a distinct increase in IC, which results in significantly larger production cost. While the potential has to be addressed by dedicated investigations in the future, a first necessary step is the consideration of a single reaction solvent along a pathway. A second attempt to reduce the IC, is a **simultaneous capacity and supply chain optimization** in order to benefit from an economy of scale and simultaneously keep the biomass transportation cost to a minimum (cf. Section 5.1.7.2).

In addition, energy integration is of vital importance for improving the sustainability of biorefineries and results in a direct reduction of greenhouse gas emissions. While the performed analyses of the heat integration potential in Sections 5.1.6 and 5.2 reveal only minor energy reduction potential, an extended analysis should take into account additional **energy integration techniques**. Heat pumps might improve the potential for heat integration, especially in case of small temperature differences between hot and cold process streams. Furthermore, mechanical pressure exchange systems should be considered for pressure-driven membrane processes, if nanofiltration or reverse osmosis processes are integrated in the separation system.

As highlighted by the multi-product analysis with **co-production of value-added chemicals** and fuel in Section 5.3, already small quantities of value-added chemicals may suffice to reach overall profitability of such mixed-product plants. In addition, the co-production of chemicals allows for efficient biorefineries nowadays, such that these can be used for the transition phase from conventional to renewable resources with respect to fuel production. However, the products need to be carefully selected to consider the production of noticeable market shares, market developments and

6 Biorefinery improvement potential

potential price changes, which arise from the introduction of a new product source. Also taking into account region-specific demand information, value-added products should be selected, which are required in the vicinity of the production site in order to avoid long distance transportation.

In this thesis, only biochemical chemical conversions of lignocellulosic biomass are analyzed. In order to determine optimal processes and benchmark various process concepts, the consideration of additional **alternative feedstocks and technologies** is important. With the use of the PNFA, an analysis covering all different feedstocks (e.g., starchy and woody biomass, waste residues, algaes, carbon dioxide and hydrogen) as well as technologies (e.g., biochemical, thermochemical conversion of biomass, water electrolysis using electricity from wind power), allows an overall and fair benchmark of the various processes and can identify benefits from the integration of different concepts.

Similar to classical chemical plants, but even more important, region-specific factors, such as availability and prices of feedstocks, utilities and work force, need to be considered for the design of biorefineries. In this thesis, only biochemical conversions of lignocellulosic biomass are analyzed. However, the choice of feedstock is of significant importance for the production process and the feedstock cost and availability. For Germany, which is considered in the current thesis, different regions show significant differences. While there is a high biomass availability in the south of Germany, northern Germany exhibits an excess of wind power, while intermediate regions, e.g. NRW, can take advantage of biomass and excess energy. Consequently, there will be no general solution, but it is rather of key importance to determine dedicated **region-specific biorefinery solutions**, subject to the local constraints.

One possibility to address parts of the highlighted challenges, are biotechnological conversions. Efficient microorganisms cannot only achieve a high product selectivity, but also combine and integrate multiple conversion steps. Thus, a single fermentation step can replace multiple reactors which are otherwise required for chemical conversions. This reduces the investment cost and can simplify the necessary downstream processing. Furthermore, fermentations work at (nearly) room conditions and do not require expensive catalysts. In order to further elucidate the potential of biotechnological processes, a dedicated discussion is presented in the subsequent section.

6.2 Potential of biotechnology conversions

In biorefinery concepts, a large number of platform chemicals are obtained via fermentation. These products can either be directly used as fuel (e.g., ethanol, iso-butanol) or are further converted to the final products. For this purpose, an efficient screening of fermentation and downstream processing is the key to identify sustainable and economical processes (Darkwah et al., 2018).

Figure 6.1 presents an overview of fermentation products, which can be obtained from sugar fermentation (Werpy and Petersen, 2004; Patel et al., 2006; Hermann and Patel, 2007; Dugar and Stephanopoulos, 2011; Straathof, 2014). These include alcohols, diols, ketones, lactones and carboxylic acids. Not all syntheses are successfully proven yet (Straathof, 2014), such that these are not considered in the analysis. Furthermore, non-selective fermentations (e.g., ABE), proof-of-principle fermentations with a final titer lower than 20 g/L and incomplete data-sets (acrylic and methacrylic acid) are excluded from further consideration. Products with an oxygen to carbon ratio larger than one (citric and gluconic acid) are neglected as well, since in fuel production a lower oxygen to carbon ratio compared to sugars is desired (Dahmen and Marquardt, 2016). In a first step, only a subset of the remaining 19 fermentation products are considered to demonstrate the potential of the analysis. The products considered in the analysis vary in their yield, productivity, titer, pH and aeration. In future, the analysis can be easily completed for the remaining products. Since the effect of different downstream alternatives is analyzed, an enumeration instead of an optimization is conducted, which is implemented in Matlab R2016a (MathWorks, 2016). All required data are given in Appendix C.9.

Directed evolution or genetic engineering can improve selectivity and conversion rate of microorganisms. However, as raw materials are typically the main cost-driver in fermentations (Klein-Marcuschamer and Blanch, 2013; Straathof, 2014), the yield is a key parameter for viable processing. Any improvement thereof is limited by the theoretical yield due to the overall stoichiometry of the fermentation. In order to establish a robust ranking, both the theoretical yield and the maximum yield reported in literature are utilized, which reveals the actual research status and any future improvement potential. In contrast to simple stoichiometric balances as proposed by Straathof (2014), the theoretical yields are calculated by analyzing feasible metabolic pathways (Ebert, 2014). The resulting raw material costs to provide unit energy are shown in Figure 6.2. As benchmark for the production of biofuels serves a recent E10 price as a state-of-the-art fuel.

6 Biorefinery improvement potential

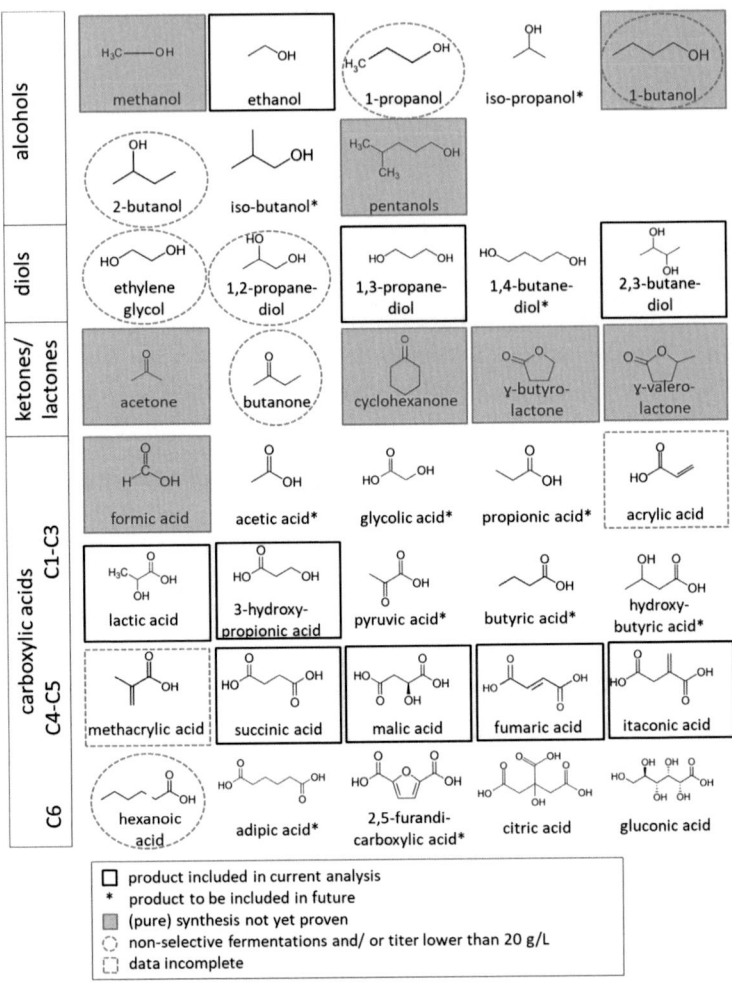

Figure 6.1: Overview of conceivable fermentation products.

A large cost discrepancy between the theoretical and maximum reported yield, indicates a need for further improvement. This is true for the fermentations to itaconic acid, 3-hydroxypropionic acid (HPA) and malic acid.

6.2 Potential of biotechnology conversions

In general, the worst performance is shown by HPA. The lowest costs can be observed with ethanol, 2,3-butanediol, 1,3-propanediol, succinic and lactic acid, all achieving a yield equal or very close to the theoretical maximum. From these promising products, 1,3-propanediol seems to offer the largest potential for further yield improvement. While ethanol is established as fuel, the other four are only platform molecules requiring a subsequent conversion into biofuels. For a competitiveness to E10 prices and, thus, a viable production of future biofuels, a sugar price of less than 0.33 $/kg is required, which is in line with recent literature (Caspeta and Nielsen, 2013).

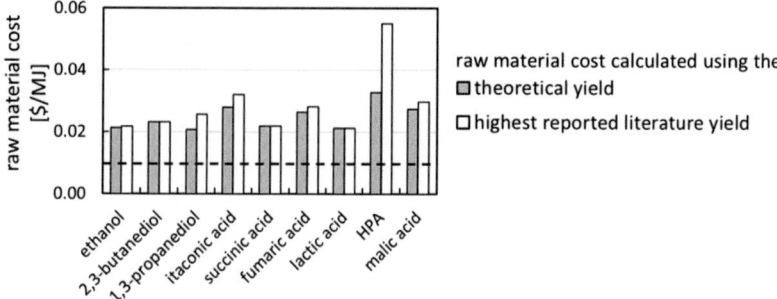

Figure 6.2: Raw material cost analysis based on the highest reported and theoretical yield. The dashed line indicates a recent E10 price in Germany as benchmark.

6.2.1 Process screening

The purification of fermentation broths is challenging, due to the presence of non-volatile products, which are highly diluted in water. The products form azeotropes with water (e.g., ethanol, ABE) or are thermally unstable (e.g., itaconic acid, citric acid), which increases the purification complexity. Furthermore, depending on the pH value, carboxylic acids dissociate, which is an undesired reaction. In addition, the separation of unconverted reactants, side products and nutrients further complicate the purification (López-Garzón and Straathof, 2014; Straathof, 2014). Thus, often fermentation setups reporting a high titer are preferred to reduce the purification effort.

6 Biorefinery improvement potential

The following analysis demonstrates the importance of considering all key reaction parameters (yield, titer, productivity, pH, aeration) simultaneously. While the yield influences the raw material cost and the titer the downstream processing, the productivity is inversely responsible for the capital cost of the fermenters (Klein-Marcuschamer and Blanch, 2013) and the energy demand for agitation and aeration. The pH value is crucial, as a subsequent adaptation might be necessary to gain carboxylic acids in their non-dissociated form. This step can significantly contribute to the production cost, as pointed out by Sauer et al. (2008) for lactic acid production.

Figure 6.3 presents the results taking into account the effort in terms of energy demand and cost for the fermentation step as well as for a subsequent downstream processing depending on the key performance indicator. For the analysis of each component, multiple datasets from literature are considered, which report the maximal yield, productivity or titer, whereas the pH value and the aeration is component-specific and, thus, often similar in the diverse literature reports (cf. Table C.18).

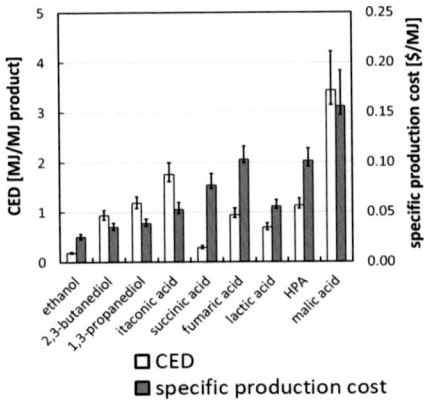

Figure 6.3: Screening results of fermentation products demonstrating the effort for fermentation and downstream processing.

Overall, the ranking of production processes is maintained. Considering a minimization of the energy demand, succinic and lactic acid are identified as promising beyond ethanol. A minimization of production cost demonstrates the production of 2,3-butanediol and 1,3-propanediol almost competitive with ethanol. Summarizing the characteristics of these promising processes, anaerobic processes with a titer larger than 135 g/L and a productivity larger than 1.8 g/L/h are preferred.

Although HPA fermentation is conducted anaerobically, a low titer causes a high effort in downstream processing. Therefore, an improvement of fermentation yield and titer are more important for the future than the improvement of the productivity. The production of malic acid is less promising. The major reasons are an aerobic fermentation along with a low productivity (0.94 g/L/h). Here a productivity increase is most promising for future improvement. In case the productivity might be limited by the oxygen transfer rate, alternatives are the utilization of pure oxygen instead of air and an in-situ product removal (Straathof, 2014).

A similar example is the production of itaconic acid. Here, this analysis revealed a major bottleneck caused by the electricity requirement for agitation and aeration during fermentation along with a low productivity. As a consequence, attempts to reduce the aeration effort for the production of itaconic acid have been enabled leading to a successful reduction of the specific energy demand during fermentation by a factor of six (Kreyenschulte et al., 2016). With the help of a systematic performance evaluation, bottlenecks and thereby specific improvement options are identified and tailor-made for the fermentation product and microorganism.

To analyze the different downstream processing alternatives and their influence on the energy demand and production cost, the production of itaconic and succinic acid are described in more detail. For this purpose, all data-sets from literature for the fermentation to both acids are considered (cf. Tables C.19-C.20), thereby covering the full performance range.

Figure 6.4 shows these ranges for all data-sets of both acids (itaconic acid Figure 6.4 a) and succinic acid Figure 6.4 b)), visualizing a broad distribution. The energy demand and cost vary by a factor of more than 100 between the best and worst performing setup of the acids. While this illustrates the improvement potential of proof-of-principle compared to optimized data-sets, it also stresses the importance of analyzing the process performance as early as possible and thereby supporting the development.

The downstream processes for both acids show similar effects. In order to gain the acids in their non-dissociated form, a precipitation (Pre) causes higher cost, while an electrodialysis (ED) has a higher energy demand. The reason for higher production cost using a precipitation instead of an electrodialysis, is a stoichiometric production of gypsum. Since the gypsum is contaminated with nutrients from fermentation, the gypsum needs to be disposed and cannot be sold (López-Garzón and Straathof, 2014). Therefore, precipitation causes a significant waste production. Furthermore, the requirement of a precipitation agent (calcium hydroxide, calcium oxide) raises the production cost.

6 Biorefinery improvement potential

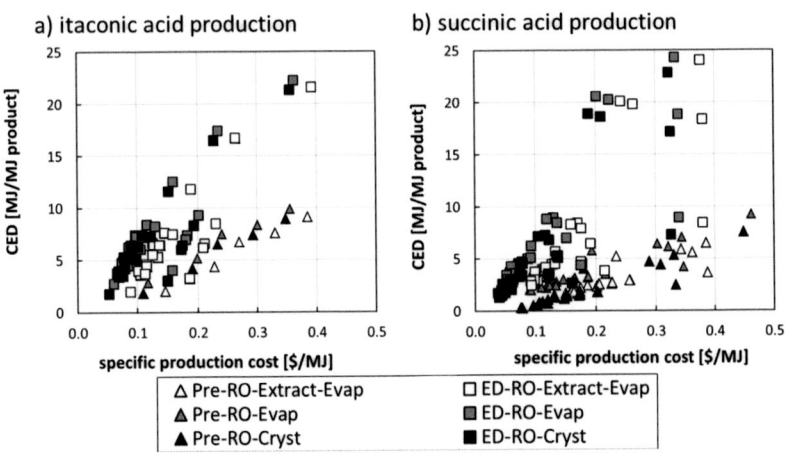

Figure 6.4: Results of all data-sets for the various downstream processing alternatives in the production to itaconic acid (a) and succinic acid (b).

In the production process to itaconic acid, an electrodialysis is preferred both in terms of energy demand and production cost. The reason is a low fermentation pH of 3 (Hevekerl et al., 2014) leading to a low dissociation degree, such that only a minor influence of the pH adaptation step is detected. In addition, product losses occur during precipitation caused by a residual solubility of the calcium salt (López-Garzón and Straathof, 2014). These losses are the reason for a superior CED using electrodialysis compared to precipitation.

In case of succinic acid production, the pH value of all fermentations is in between 6 and 7.5, such that succinate is formed in all cases. Therefore, the influence of the pH adaptation step is higher compared to itaconic acid fermentation.

The preferred final purification for both acids consists of a reverse osmosis (RO) connected to a cooling crystallization (Cryst). Herein, the reverse osmosis increases the product concentration. The subsequent cooling crystallization takes advantage of a temperature-dependent solubility of the acids in water. A recycle from crystallization to preheating enhances the overall downstream yield (cf. Figure C.4). Compared to crystallization, a final thermal separation (Evap) raises the energy demand significantly. While the energy duty is lowered using an extraction (Extract) prior to the thermal separation, the additional solvent raises the production cost.

6.2.2 Future improvement potential

Often only the final parameter values of titer, (averaged) productivity and yield are published. In contrast to fixed fermentation parameter values, the consideration of a fermentation course provides further improvement potential. Especially for aerobic fermentations, the significant impact of the electricity requirements during fermentation is a major challenge. Thus, attempts to increase productivity instead of the titer are more promising. This is demonstrated in Figure 6.5 for an itaconic acid fermentation (Kuenz et al., 2012) and a subsequent conversion into the fuel 3-methyltetrahydrofuran according to Geilen et al. (2010).

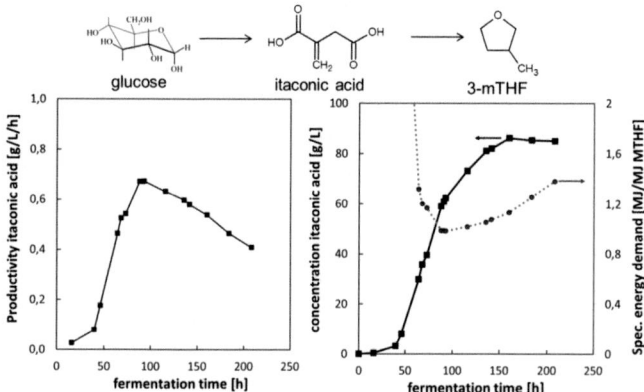

Figure 6.5: Influence of titer and productivity development during itaconic acid fermentation on the overall energy requirement for the production of 3-methyletrahydrofuran. Herein, the fermentation data of Kuenz et al. (2012) is utilized.

A maximal average productivity of 0.67 g/L/h is achieved after 93 hours, whereas the overall average productivity is 0.54 g/L/h. While a final titer of 87 g/L is achieved after 160 hours, the titer at 93 hours is 62 g/L. Nevertheless, the specific energy demand shows a minimum at the point of maximal average productivity instead of the point of maximal titer. Thus, stopping the fermentation earlier leads to a cost reduction by 15% (not explicitly shown in Figure 6.5) and an energy reduction by 16%. Hence, an analysis based on the fermentation course opens up further improvement potential. While discrete parameter values are utilized herein, even fermentation kinetics can be integrated in the future.

6 Biorefinery improvement potential

6.2.3 Summary

The analysis of fermentation processes demonstrates that the processes considered are not competitive to ethanol. Beyond ethanol, promising platform chemicals are identified, which can be either produced in a cost-competitive (2,3-butanediol, 1,2-propanediol) or a CED-competitive (succinic and lactic acid) manner. This requires a highly efficient subsequent conversion into fuels to maintain competitiveness with ethanol. While ethanol can be produced from both sugar fractions, the platform chemicals are solely obtained from the hexose fraction. Thus, the pentose fraction can be utilized for the co-production of value-added chemicals.

The process analysis of all available itaconic and succinic acid fermentation data-sets illustrates the large improvement potential of initial proof-of-principle experiments (high cost and energy demand). The superior fermentation data-sets are way more cost efficient and require less energy. This is also interesting for novel fermentations, which are right now only capable of producing small quantities. With genetic engineering or directed evolution, these might play an important role in the future.

Additional improvement potential is enabled by considering all key fermentation influence factors (yield, titer, productivity, pH and aeration) simultaneously. The example of itaconic acid demonstrates that it might be useful to stop an aerobic fermentation earlier to obtain a high productivity rather than a high titer.

The case study of biotechnological conversions illustrates the importance of an early-stage screening method like the PNFA coupled to genetic engineering approaches to guide future research developments. In the following chapter, the results of this thesis are summarized, major conclusions are drawn and discussed and a brief outlook on future methods and case study improvements are given.

Chapter 7

Conclusions and outlook

Due to the envisaged feedstock change from conventional to renewable resources, a high number of novel processes are proposed in literature. In order to evaluate these pathways, screening methods are required for an initial assessment of their sustainability and economic efficiency. Existing methods (e.g., Bao et al., 2011; Santibanez-Aguilar et al., 2011; Rangarajan et al., 2014a; Cheali et al., 2014; Kelloway and Daoutidis, 2014; Zondervan et al., 2011; Celebi et al., 2017) are based on process data from literature or tedious simulation studies. These simulation studies often lack the required robustness and the results depend on user-specified design decisions. Hence, the scope of this thesis is the development of a systematic screening method for a fast evaluation of existing and novel processes at an early design stage. The RNFA proposed by Voll and Marquardt (2012a) is a first step towards a systematic evaluation of biorefinery reaction pathways, which evaluates the mass efficiency of reaction pathways assuming ideal separations.

Herein, the method is extended towards the PNFA. The PNFA additionally evaluates the process performances and systematically addresses the choice, feasibility and effort of separations independent of design decisions. Thereby, a fast evaluation of the economic efficiency and sustainability of a high number of processing pathways is enabled. Thus, the PNFA fits into the proposed biorefinery process framework, which gradually increases the model complexity from RNFA to PNFA and conceptual design, while the number of alternative pathways is decreased. The PNFA is completed by a heat integration potential analysis, a consideration of the biomass supply chain and the selection of an optimal product portfolio based on varying prices.

7 Conclusions and outlook

Conclusions

The process results obtained with the PNFA fit to the literature values. However, the comparison also reveals large uncertainties for a specific process, which is caused by different available process and price parameters, selected feedstocks and technologies. Therefore, comparability is in general difficult. The comparison with RNFA and conceptual design demonstrates the functionality of the biorefinery process framework. Compared with the RNFA, the PNFA assesses the process energy demand. The comparison with a conceptual design study reveals the enzyme cost as missing parameter in the PNFA. While in principle, an enzyme price factor can be easily introduced in the model, a reliable estimation of enzyme prices is challenging, since these vary substantially and are also expected to decrease in the future (Klein-Marcuschamer et al., 2012; Johnson, 2016; Valdivia et al., 2016).

The PNFA is applied on case studies to produce single-product fuels and fuel mixtures from biomass (cf. Chapter 5). The results are based on the PNFA model and assumptions described in Chapter 3 and the processing pathways described in Chapter 5. All results identify the production to ethanol as most promising in terms of production cost, energy demand and global warming potential, such that no superior alternative novel fuel or fuel mixture can be identified. The ethanol process is superior, because it is well studied, it utilizes both the hexose and pentose fraction, it requires a single conversion step from sugars to the product and the reaction is conducted in water, such that an expensive water removal is postponed to the product purification. Even though the mixture of ethanol and water exhibits an azeotrope, the separation is straightforward, such that the process energy demand is low.

In contrast, the novel processes mainly start from a single sugar fraction only, require multiple conversion steps and a solvent change from water to an organic solvent. These solvent changes as well as final purifications are often expensive since the components exhibit boiling points larger than the boiling point of water or the respective solvent and form azeotropes. This often leads to energy-intensive separation schemes, where the high-boiling product needs to be removed in a first step.

Beyond ethanol production, the process to iso-butanol is most promising. The reasons are a similar biotechnological one-step conversion from sugars and a miscibility gap between water and iso-butanol, which can be efficiently exploited for purification. In contrast to ethanol production, the pentose fraction is not valorized. Iso-butanol is a promising future fuel with high octane number, superior higher energy density, reduced volatile emissions and less corrosion tendency compared to ethanol (Fenkl et al., 2016; British Petroleum, 2018; The DOW Chemical Company, 2018).

A potential valorization of the pentose fraction can be the conversion into ethanol to obtain a mixture with iso-butanol. While no publications are available on the analysis of such a mixture, both products can be blended to gasoline (Fenkl et al., 2016). However, the mixture is only useful if a superior performance in the engine is obtained compared to pure ethanol, since clearly the production of ethanol alone is more favorable. Alternatively, the pentose fraction can be valorized via furfural, which is diluted in water. The analysis in Section 5.1 identifies the thermal separation of furfural and water expensive, whereas a bi-phasic setup as described in Section 5.2 is more promising.

The benefit of utilizing both sugar fractions to obtain fuel mixtures is demonstrated once more in Section 5.2 for the co-production of DNBE and a C8-C16 mixture for a compression-ignition engine. Herein, the co-production is superior compared to DNBE production only. Nevertheless, the resulting mixture is not competitive to ethanol due to a high number of conversion steps.

When considering biomass supply chain in addition to the process costs, none of the analyzed fuel processes including ethanol production, is profitable. The reasons are higher feedstock costs compared to first generation biofuels, an expensive feedstock transportation compared to conventional fuels and a low product value. Therefore, the transition from conventional and first generation biofuels to fuels derived from lignocellulosic biomass, requires either substantial tax advantages or the co-production of value-added chemicals to obtain profitability (cf. Section 5.3). A biorefinery co-production of fuel (ethanol) and chemicals (iso-butanol, 2,3-butanediol and furfuryl alcohol) reveals a satisfying profitability and only minor fractions of chemicals are required to break-even. Therefore, even a biorefinery with few product streams allows for an efficient fuel production.

The key results of the case studies are condensed into (general) improvement factors for biorefineries. These key factors are the valorization of both sugar fractions, efficient solvent, in particular, water removal, reduction of investment cost, realisation of one-pot reactions, a simultaneous capacity and supply chain optimization, the consideration of alternative energy integration techniques, a co-production of value-added chemicals and the integration of additional resources, which ultimately leads to region-specific biorefinery solutions.

A potential solution, which addresses multiple of these improvement factors, are biotechnological conversions. Biotechnological conversions are conducted in water, require a single conversion step from sugars only and are often very selective. The analysis identifies again ethanol as best performing process.

7 Conclusions and outlook

Beyond ethanol, 2,3-butanediol, 1,2-propanediol are promising from a cost perspective and succinic and lactic acid from an energy demand perspective. However, all of these processes have in common that only the hexose fraction is utilized and that platform chemicals are produced instead of final fuels. This requires an efficient conversion into the final fuels in order to maintain competitiveness. Furthermore, the analysis illustrates the improvement potential of initial proof-of-principle to optimized fermentation setups as well as the importance of a simultaneous consideration of all fermentation key characteristics, i.e., yield, productivity, titer, pH and aeration.

Outlook

The PNFA can be further improved to extend the analysis opportunities. Since the choice of a reaction solvent influences not only the reaction, but also the subsequent separation, a connection to a solvent screening allows the identification of optimal solvents. For this purpose, the pathways with expensive separation steps are selected and the solvent screening is used to identify a suitable solvent, which allows an easier separation. The solvent then needs to be applied to the reaction system in order to determine the reaction yield and selectivity.

For the solvent screening, different methods are available in literature, which are either based on group-contribution approaches (Pretel et al., 1994; Gani et al., 2005) or COSMO-RS (Eckert and Klamt, 2002; Scheffczyk et al., 2016) considering solvent properties like selectivity or phase distribution (Pretel et al., 1994) or the process performance (Papadopoulos and Linke, 2006; Kossack et al., 2008; Scheffczyk et al., 2016). Even though any solvent selection method can be used and connected to the PNFA, the approaches of Kossack et al. (2008) and Scheffczyk et al. (2016) are the most promising methods to use as these rely on the same thermodynamically-sound separation models applied in the PNFA. Therefore, the solvent screening directly provides the input data required for the PNFA.

In order to address the complex separations and further reduce the separation effort, alternative separation technologies and process intensification techniques need to be analyzed. These strategies need to be applicable to non-ideal systems, e.g., systems exhibiting an azeotrope and need to be further integrated into screening approaches like the PNFA.

Especially for the coupling of PNFA to genetic engineering, the introduction of additional separation and purification strategies is important for a profound process analysis. In order to examine the performance of intensified separation techniques, different attempts have been recently introduced in literature (e.g., Kuhlmann and Skiborowski (2017); Demirel et al. (2017) and Tula et al. (2017)). These approaches rely on a decomposition of a separation task into phenomena building blocks to identify suitable intensified separations. Since the problem size scales with the number of alternative (intensified) separation techniques, it is not applicable for all available separation tasks in the PNFA, but can rather be used as debottlenecking approach.

Since a variety of potential products and processes are conceivable, an integration of product and process design supports the development of an overall optimum from a product and process perspective. Various attempts for a simultaneous product and process design exist in literature (e.g., Marvin et al., 2013a; Ng et al., 2015; Hashim et al., 2017). Hechinger et al. (2010) proposed an integration of RNFA to product design, which was conducted by Dahmen and Marquardt (2016). Since the PNFA facilitates a more detailed process analysis, an extension of the approach of Dahmen and Marquardt (2016) to an integration of PNFA and product design is beneficial.

For these purposes, an automated data assembly for the PNFA is necessary. This includes the setup of reaction networks, the systematic identification of separation steps and the assessment of separation performance in terms of auxiliary or energy demand. Once data assembly is automated, an integration with ontology-based repositories as proposed by Siougkrou and Kokossis (2016a) or Bertran et al. (2017) allows a consistency check with other research groups. More importantly, it will provide novel data on processing steps and pathways, which can be included in the repository and as such reviewed and used by any research group.

In the future, the effect of uncertainties needs to be further analyzed. While herein the influence of individual parameters are considered, an analysis of the interdependence of parameters is needed to obtain more robust conclusions. An example are uncertainties in the reaction yield, e.g., due to scale-up, which influence the subsequent separation task. This extended analysis will help to identify the most important parameters and process steps, which can then be evaluated in more detail.

PNFA has been applied for biochemical conversion of lignocellulosic biomass. In the future, additional renewable feedstocks, hydrogen generated using water electrolysis powered by wind or solar energy, carbon dioxide as well as alternative conversion techniques like thermochemical approaches need to be included to determine a region-specific feedstock and process optimum. A first attempt to analyze alternative feedstocks like hydrogen or carbon dioxide using the RNFA shows an economic drawback compared to biomass conversions due to currently high costs for renewable hydrogen (Monigatti, 2016). Since this is region-specific and might change, a case study covering all these aspects, enables the determination of optimal plants.

Appendix A

Separation models

In the following, the additional separation models required for the purification of fermentation broths are briefly described. The fermentation itself is accessed by the investment cost for the fermenters, the utility requirement for agitation and aeration as well as the raw material cost. All required parameters are given in Appendix C.1. Every separation step is based on an incoming flux f_{IN} either directly from fermentation or from a previous separation step. For all separations, the outgoing product flux $f_{product,OUT}$ is described. Possible interconnections between different separations are shown in Figure C.2.

Precipitation The model for precipitation is exemplary shown for calcium hydroxide $(Ca(OH)_2)$ as precipitation agent. An alternative is the utilization of calcium oxide (López-Garzón and Straathof, 2014) with a similar stoichiometry. For a dissociated di-protonic acid $(H_2^+A^{2-})$, the general reaction equations read as follows

$$H_2^+A^{2-} + Ca(OH)_2 \rightarrow CaA \downarrow + 2H_2O \quad (A.1)$$
$$CaA + H_2SO_{4,aq} \rightarrow CaSO_4 \downarrow + H_2A \quad (A.2)$$

In a first step, calcium salt (CaA) and water are formed. The calcium salt reacts with diluted sulphuric acid $(H_2SO_{4,aq})$ to calcium sulfate $(CaSO_4)$ and a non-dissociated acid form (H_2A). In order to accurately determine the incoming mass flow, an averaged molar mass $\bar{M}$ is calculated considering for the number of dissociations N_κ the degree of dissociation κ,

$$\bar{M} = \sum_i^{N_\kappa} (\kappa_i M_i). \quad (A.3)$$

A Separation models

Herein, κ equals one for a non-dissociated molecule, whereas κ increases by one for each dissociation step. The required calcium hydroxide flux is then determined based on the incoming product flux $f_{product,IN}$, the number of acid dissociations divided by a factor of two for the number of $Ca(OH)_2$ dissociations,

$$f_{Ca(OH)2,IN} = f_{product,IN} \frac{N_{\kappa,product}}{2}. \tag{A.4}$$

Based on the assumption of a full conversion, the sulfuric acid flux equals the calcium hydroxide flux according to,

$$f_{H2SO4,IN} = f_{Ca(OH)2,IN}. \tag{A.5}$$

The waste water flux is the sum of the incoming water flux along with the produced water in the first reaction (Equation A.2),

$$f_{H2O,w} = f_{H2O,IN} + 2f_{Ca(OH)2,IN}. \tag{A.6}$$

A residual anion solubility in water prevents a full conversion in the second reaction. Based on the solubility of the anion in water Sol_{Anion}, the dissolved anion flux can be described by,

$$f_{Anion,aq} M_{Anion} = Sol_{Anion} \frac{f_{H2O,w} M_{H2O}}{\rho_{H2O}}. \tag{A.7}$$

The final product flux $f_{product,OUT}$,

$$f_{product,OUT} M_{product} = Y_{Pre}(f_{IN}\bar{M} - f_{Anion,aq} M_{Anion}), \tag{A.8}$$

is then determined on the difference between the feed flux f_{IN} and the dissolved anion flux, assuming otherwise a full yield Y_{Pre} of 1. The calcium sulfate flux, which needs to be disposed, is calculated based on the required sulphuric acid flux

$$f_{CaSO4,OUT} = f_{H2SO4,IN}. \tag{A.9}$$

The precipitation effort is then accounted for by auxiliary cost (calcium hydroxide, sulphuric acid), waste disposal cost (waste water, calcium sulfate) and product losses caused by a residual solubility of the anion in water.

Electrodialysis The effort to gain acids in their non-dissociated form using electrodialysis is considered by the specific electricity requirement $E_{spec,ED}$ for the transport over the membrane,

$$E_{elec} = E_{spec,ED} \sum_{i}^{N_\kappa} f_{\kappa_{elec},i} i. \tag{A.10}$$

In a first step, only the transport over the membrane for the dissociated species $f_{\kappa_{elec}}$ is taken into account

$$f_{\kappa_{elec,i}} = f_{product,IN}\kappa_i \quad \forall \ \kappa > 1. \tag{A.11}$$

The product flux is then described by

$$f_{product,OUT}M_{product} = Y_{ED}\sum_{i=1}^{N_\kappa} f_{\kappa_i} M_{\kappa,product}. \tag{A.12}$$

Reverse Osmosis The reverse osmosis is used to increase the product concentration by removing water. For this purpose, first the incoming water flux is determined based on the product concentration c,

$$f_{H2O,IN}M_{H2O} = \rho_{H2O}\frac{f_{product,IN}M_{product}}{c_{product,IN}}. \tag{A.13}$$

The total feed stream is then the sum of the incoming water and the product stream according to,

$$f_{IN}M_{IN} = f_{H2O,IN}M_{H2O} + f_{product,IN}M_{product}. \tag{A.14}$$

In order to determine the product retentate stream,

$$f_{product,OUT}M_{product} = Ret \cdot f_{product,IN}M_{product}, \tag{A.15}$$

the products retention Ret is required as input. Preventing a product crystallization in the RO, a safety margin Sol_{Limit} is assumed such that the product remains diluted in the outgoing retentate water stream,

$$f_{H2O,OUT}M_{H2O} = \rho_{H2O}\frac{f_{product,OUT}M_{product}}{Sol_{Limit} \cdot Sol_{product}}. \tag{A.16}$$

The outgoing permeate water stream is then calculated based on a simple mass balance. The concentration of the product in the retentate stream is then based on the quotient of the product and water stream according to

$$c_{product,OUT} = \rho_{H2O}\frac{f_{product,OUT}M_{product}}{f_{H2O,OUT}M_{H2O}}. \tag{A.17}$$

In order to increase the product solubility prior to the RO and thereby be capable of removing a larger water fraction, a heat exchanger HEX can be installed with the following specific heat requirement according to

$$E_{heat} = \frac{f_{H2O}M_{H2O}c_{pH2O} + f_{product}M_{product}c_{pProduct}}{f_{IN}}(T_{HEX} - T_{IN}). \tag{A.18}$$

A Separation models

According to Melin and Rautenbach (2007), the specific electricity requirement is based on the osmotic pressure $p_{osmotic}$ and the pump efficiency η_{pump},

$$E_{elec} = M_{IN}/\rho_{H2O} \cdot \frac{p_{osmotic}}{\eta_{pump}}. \tag{A.19}$$

For a first assessment of the osmotic pressure, Melin and Rautenbach (2007) proposes a pressure contribution from friction $p_{friction}$ and a specific RO pressure factor p_{RO}, which is multiplied with the products solubility,

$$p_{osmotic} = p_{friction} + \frac{p_{RO} R T_{IN} \cdot Sol_{product}}{M_{product}}. \tag{A.20}$$

Cooling crystallization A cooling crystallization takes advantage of a temperature-dependent solubility to separate the solid product from a mother liquor (ML). Herein, the incoming water feed equals the water of the mother liquor assuming in a first step, an ideal solid product without a residual moisture content, which is a simplification. The water feed stream is then determined according to

$$f_{H2O,IN} M_{H2O} = f_{H2O,ML} M_{H2O} = \frac{f_{product,IN} M_{product} \rho_{H2O}}{c_{product,IN}}. \tag{A.21}$$

The remaining product in the mother liquor is calculated based on the products solubility,

$$f_{product,ML} M_{product} = \frac{Sol_{product} f_{H2O,ML} M_{H2O}}{\rho_{H2O}}, \tag{A.22}$$

and can be recycled. The solid product stream is the difference between the incoming product and the product remaining in the mother liquor,

$$f_{product,OUT} = Y_{Cryst}(f_{product,IN} - f_{product,ML}). \tag{A.23}$$

The cooling water requirement is based on the temperature difference and the latent enthalpy of fusion E_{fus},

$$E_{cool} = \frac{(f_{H2O,IN} M_{H2O} cp_{H2O} + f_{product,IN} M_{product} cp_{product}) \cdot (T_{Cryst} - T_{IN}) + E_{fus}}{f_{IN}} \tag{A.24}$$

Herein, the enthalpy of fusion is based on the solid product flux,

$$E_{fus} = f_{product,solid} M_{product} h_{product,fus}. \tag{A.25}$$

Appendix B

Green chemistry metrics

The following table summarizes applicable sustainability criteria for early-stage screening. The criteria are ordered according to the green chemistry and engineering principles. Criteria marked with a star are redundant to other criteria or can be determined based on a summary of multiple criteria. The toxicity score, *ToxS*, is derived according to Uhlman and Saling (2010) and Ulonska et al. (2016b). The mass intensity criteria (Eq. B.13) considers the utilization of catalysts, auxiliaries and solvents besides the reagents.

Table B.1: Sustainability criteria for the early design stage.

	Criteria	Equation	
min. waste	E-factor	$E-factor = \frac{\sum_{i=1, i \neq water}^{N_w} b_{w,i} M_i}{\sum_{i=1}^{N_{products}} b_{product,i} M_{product,i}}$	(B.1)
	Waste intensity	$WI = \frac{\sum_{i=1, i \neq water}^{N_w} b_{w,i} M_i}{\sum_{j=1}^{N_{supply}} f_{supply,j} A_{i,j} M_i}$	(B.2)
	Waste percentage	$WP = \frac{WI}{PMI}$	(B.3)
	Waste energy ratio	$WER = \frac{\sum_{i=1, i \neq water}^{N_w} b_{w,i} M_i}{e_{heat} E_{heat} + e_{elec} E_{elec}}$	(B.4)
	Emissions	$EM = \frac{b_{CO2,r} M_{CO2} + b_{CO,r} M_{CO} + 25 b_{CH4,r} M_{CH4}}{\sum_{i=1}^{N_{products}} b_{product,i} M_{product,i}}$	(B.5)

B Green chemistry metrics

	Criteria	Equation	
max. mass efficiency	Atom economy	$AE = \dfrac{\sum\limits_{i=1}^{N_{products}} M_{product,i}}{\sum\limits_{i=1}^{N_{reagents}} M_{reagent,i}}$	(B.6)
	Atom utilization	$AU = \dfrac{\sum\limits_{i=1}^{N_{products}} b_{product,i} M_{product,i}}{\sum\limits_{i=1}^{N_{products}} b_i M_i}$	(B.7)
	Element efficiency	$EE = \dfrac{\sum\limits_{i=1}^{N_{products}} b_{product,element,i} M_{product,i}}{\sum\limits_{j=1}^{N_{reagents}} f_{j,element} A_{i,j} M_{element,i}}$	(B.8)
	Carbon efficiency	$CE = \dfrac{\sum\limits_{i=1}^{N_{products}} b_{product,carbon,i}}{\sum\limits_{j=1}^{N_{reagents}} f_{j,carbon} A_{i,j}}$	(B.9)
	Reaction mass efficiency	$RME = \dfrac{\sum\limits_{i=1}^{N_{products}} b_{product,i} M_{product,i}}{\sum\limits_{j=1}^{N_{products}} f_{supply,j} A_{i,j} M_i}$	(B.10)
	Resource consumption*	$RC = \dfrac{1}{RME}$	(B.11)
	Generalized RME*	$gRME = \dfrac{1}{E\text{-}factor+1}$	(B.12)
	Mass intensity*	$MI_i = \dfrac{\sum\limits_{j=1}^{N_{supply}} f_{supply,j} A_{i,j} M_i}{\sum\limits_{i=1}^{N_{products}} b_{product,i} M_{product,i}}$	(B.13)
	Process mass intensity*	$PMI = E\text{-}factor + 1$	(B.14)
	Optimum efficiency*	$OE = \dfrac{RME}{AE}$	(B.15)
	Space time yield	$STY = \dfrac{\sum\limits_{i=1}^{N_{products}} b_{product,i} M_{product,i}}{V_{reactor} t_{reaction}}$	(B.16)
min. $b_{hazardous}$	Toxicity Potential	$ToxP = \sum\limits_{i=1}^{N_{products}} b_{product,i} M_{product,i} ToxS_{product}$	(B.17)
	Effective mass yield	$EMS = \dfrac{\sum\limits_{i=1}^{N_{products}} b_{product,i} M_{product,i}}{\sum\limits_{j=1}^{NC} f_{non-benign\ reagents,j} A_{i,j} M_i}$	(B.18)

	Criteria	Equation	
min. $f_{auxiliaries}$	Hydrogen requirement	$H2 = \dfrac{f_{H2,supply} M_{H2}}{\sum_{i=1}^{N_{products}} b_{product,i} M_{product,i}}$	(B.19)
	Solvent intensity	$SI = \dfrac{\sum_{j=1, j \in solvents}^{N_{supply\ reactions}} f_{1,j} A_{i,j} M_i}{\sum_{j=1, j \neq solvent}^{N_{supply\ reactions}} f_{1,j} A_{i,j} M_i}$	(B.20)
	Solvent recovery (energy)	$SR = \dfrac{E_{solvent\ recovery}}{\sum_{i=1}^{N_{products}} b_{product,i} M_{product,i}}$	(B.21)
	Solvent energy ratio	$SER = \dfrac{E_{Solvent,recovery}}{e_{heat} E_{heat} + e_{elec} E_{elec}}$	(B.22)
min. energy	Energy efficiency combustion	$EC = \dfrac{\sum_{i=1}^{N_{products}} b_{product,i} H_{comb,i}}{\sum_{j=1}^{N_{supply\ reactions}} f_{1,j} A_{i,j} H_{comb,i}}$	(B.23)
	Energy efficiency formation	$Eform = \dfrac{\sum_{i=1}^{N_{products}} b_{product,i} H_{form,i}}{\sum_{j=1}^{N_{supply\ reactions}} f_{1,j} A_{i,j} H_{form,i}}$	(B.24)
max. f_{bio}	Renewables intensity	$RI = \dfrac{\sum_{j=1, j \in renewable}^{N_{supply\ reactions}} f_{1,j} A_{i,j} M_i}{\sum_{i=1}^{N_{products}} b_{product,i} M_{product,i}}$	(B.25)
	Renewables percentage*	$RP = \dfrac{RI}{PMI}$	(B.26)

Appendix C

Case study parameters

This chapter presents all required case study parameters as well as additional information and results, which are not included in the main text for brevity reasons. The chapter starts in Section C.1 with an overview of all general parameters. In Section C.2 the reaction parameters of the reaction network for the case study in Section 5.1 are given. The properties, which are required for both the RNFA and PNFA are presented in Section C.4. Section C.5 then summarizes all active reactions and separations for the RNFA and PNFA results presented in Section 5.1. Additional information on the complementary supply chain design is described in Section C.6 and on the multi-product biorefinery in Section C.7. The final mixture compositions derived in the two co-production scenarios described in Section 5.2 are given in Section C.8. Since the fermentation screening requires additional information on the respective fermentation setups as well as on the product properties, these are presented in Section C.9 along with an exemplary flowsheet for the final purification of itaconic acid. This chapter ends with a summary of additional information required for the conceptual design case study of Section 4.2.1.

C.1 Overview parameters

In Table C.1 all model and cost parameters and their respective references are given, which are required utilized in this thesis for the RNFA and PNFA of the various case studies.

C.1 Overview parameters

Table C.1: Model input parameter

Parameter	Unit	Value	Reference
reference year	-	2014	Ulonska et al. (2016a)
biomass cost	$\frac{\$}{kg}$	0.05	Kamm et al. (2006)
hydrogen cost	$\frac{\$}{kg}$	2.80	Ruth (2011)
water	$\frac{\$}{kg}$	0.0005	Sinnott et al. (2005)
dimethylsulfoxid	$\frac{\$}{kg}$	1.475	Anhui Jinao Chemical Co. Ltd. (2016)
2-methyltetrahydrofuran	$\frac{\$}{kg}$	7.5	Wuhan Benjamin Pharmaceutical Chemical Co. Ltd (2016)
methylenechlorid	$\frac{\$}{kg}$	0.55	Shanghai Polymet Commodities Ltd. (2016)
γ-butyrolactone	$\frac{\$}{kg}$	1.417	Zauba Technologies and Data Services Pvt Ltd. (2015)
1-4-dioxane	$\frac{\$}{kg}$	2.695	Jiangsu Senxuan Pharmaceutical And Chemical Co. Ltd. (2016)
calcium hydroxide	$\frac{\$}{kg}$	0.09	Sinnott et al. (2005)
calcium sulfate disposal	$\frac{\$}{kg}$	0.092	López-Garzón and Straathof (2014)
glucose/xylose	$\frac{\$}{kg}$	0.33	Viell et al. (2013)
glycerol	$\frac{\$}{kg}$	0.47	Viell et al. (2013)
$IC_{fermenter}$	$\frac{\text{Mio. \$}}{kg}$	1.35	Kumar et al. (2012)
$V_{fermenter}$	m^3	500	Kumar et al. (2012)
interest rate ir	-	0.08	-
plant runtime n	-	10	-
CEPCI (2010)	-	550.8	Lozowski (2015)
CEPCI (2014)	-	576.1	Lozowski (2015)
Inv1	-	7000	El-Halwagi (2012)
Inv2	-	0.68	El-Halwagi (2012)
waste	$\frac{\$}{t}$	28.86	Humbird et al. (2012)
steam	$\frac{\$}{t}$	9.5	El-Halwagi (2012)
electricity	$\frac{\$}{kWh}$	0.075	El-Halwagi (2012)
cooling water	$\frac{\$}{kg}$	0.065	El-Halwagi (2012)
refrigerant	$\frac{\$}{\text{MM BTU}}$	35	El-Halwagi (2012)
e_{Heat} (german industry, 2010)	-	1.12	International Institute for Sustainability Analysis and Strategy (2016)
e_{Elec} (german industry, 2010)	-	2.28	International Institute for Sustainability Analysis and Strategy (2016)
gwp_{Heat} (german industry, 2010)	$\frac{g\ CO_{2,eq.}}{MJ}$	72	International Institute for Sustainability Analysis and Strategy (2016)
gwp_{Elec} (german industry, 2010)	$\frac{g\ CO_{2,eq.}}{MJ}$	171	International Institute for Sustainability Analysis and Strategy (2016)

C Case study parameters

Parameter	Unit	Value	Reference
$M_{biomass}$	$\frac{g}{mole}$	162	Voll and Marquardt (2012b)
lower bound cellulose	-	0.4	Huber et al. (2006)
lower bound hemicellulose	-	0.16	Huber et al. (2006)
lower bound lignin	-	0.09	Huber et al. (2006)
upper bound cellulose	-	0.75	Huber et al. (2006)
upper bound hemicellulose	-	0.32	Huber et al. (2006)
upper bound lignin	-	0.23	Huber et al. (2006)
P_{fuel}	$\frac{Euro}{L}$	0.312	Aral Service (2016)
$P_{chem,weakresponse}$	-	50	factor of 2 for q=1
$P_{chem,strongresponse}$	-	90	factor of 10 for q=1
$P_{biomass,weakresponse}$	-	50	factor of 2 for q=1
$P_{biomass,strongresponse}$	-	450	factor of 10 for q=1
$E_{elec,fermentation,aeration}$	$\frac{kW}{m^3}$	2.5	Hermann et al. (2007)
$E_{elec,fermentation,agitation}$	$\frac{kW}{m^3}$	0.5	Hermann et al. (2007)
$E_{spec,ED}$	kWh/equivalent	0.1	Hermann et al. (2007)
P_{RO}	bar	1.1	Melin and Rautenbach (2007)
$P_{friction}$	bar	5	Melin and Rautenbach (2007)
$\eta_{pump,RO}$	-	0.8	Melin and Rautenbach (2007)
Ret	-	0.995	Assumption to prevent crystallization in RO
Sol_{Limit}	$\frac{kg}{m^3}$	0.95	López-Garzón and Straathof (2014)
Sol_{Anion}	$\frac{kg}{m^3}$	12.3	Verkehrsrundschau (2016)
VR_{2011}	-	119.16	Verkehrsrundschau (2016)
VR_{2014}	-	120.385	
Producer index (averaged) 2006	-	212.4	Federal Reserve Bank St. Louis (2017)
Producer index (averaged) 2012	-	306.9	Federal Reserve Bank St. Louis (2017)
Producer index (averaged) 2014	-	289.0	Federal Reserve Bank St. Louis (2017)
Producer index (averaged) 2016	-	227.3	Federal Reserve Bank St. Louis (2017)
exchange rate	$\frac{Euro}{US\$}$	0.9206	SIX Financial Information (2016)
wf EL (RNFA)	$\frac{kg}{MJ}$	0.016	Ulonska et al. (2016b)
wf Em (RNFA)	$\frac{kgCO2eq.}{}$	0.261	Ulonska et al. (2016b)
wf RC (RNFA)	-	0.090	Ulonska et al. (2016b)
wf Tox (RNFA)	$\frac{year}{kg}$	$8.35 \cdot 10^{-6}$	Ulonska et al. (2016b)

C.2 Reaction network

Table C.2 summarizes the input data as well as the references thereof required for the presented case study in Chapter 5. The network is set up using the synthesis planner of the online database Reaxys Elsevier Information Systems GmbH (2015). In addition to the main reaction data, the number of database entries within Reaxys is given identifying well-known as well as new or less studied reactions. Well-known reactions are based on extensive research and are identified by a high number of database entries ($\# \geq 100$). New reactions or publications based on proof-of-principles only exhibit a lower number of database entries ($\# \leq 10$). In addition, the number of database entries with a yield equal to or higher than 70% and 95% is given. In most cases intensively studied reactions have a higher number of these entries compared to new, less-known reactions. However, in cases as e.g., reaction R29 a high number of studies has been conducted, while only few succeeded in exhibiting high yields. There might be several reasons for this, like equilibrium limitations, the choice of a catalyst or even a low relevance of this reaction until now. Reaction data-sets without explicit statement of the reaction pressure are marked with "-" in Table C.2 and any contribution from a pressure change to the energy requirement of the reactions is not considered in a first step. For biomass pretreatment, fairly optimistic yield parameters are assumed, as proposed by Voll and Marquardt (2012b). A discussion thereof is conducted in Chapter 4. Since no representative reference was found for R4, same conditions as for cellulose depolymerization are assumed.

C Case study parameters

Table C.2: Input for the reaction network.

Reaction	Y [mol/mol]	Solvent	p [bar]	T [°C]	#Reaxys $Y \geq 0.95$	#Reaxys $Y \geq 0.70$	#Reaxys	Reference
R1	0.97	-	-	-	-	-	-	Voll and Marquardt (2012b)
R2	0.97	-	-	-	-	-	-	Voll and Marquardt (2012b)
R3	0.97	-	-	-	-	-	-	Voll and Marquardt (2012b)
R4	0.90	water	1	50	-	-	-	Manonmani and Sreekantiah (1987)
R5	0.90	water	1	50	2	6	41	Manonmani and Sreekantiah (1987)
R6	0.95	water	1	32	0	0	13	Humbird et al. (2012)
R7	0.85	water	1	32	0	0	6	Humbird et al. (2012)
R8	0.70	water	32	238	0	5	109	Kamm et al. (2006)
R9	1.00	water	18	208	2	3	24	Kamm et al. (2006)
R10	0.97	water	-	25	2	2	38	Bartoli et al. (2007)
R11	0.94	water	44	130	0	0	78	Neidleman et al. (1981)
R12	0.99	DMSO	1	100	40	43	133	Szmant and Chundury (1981)
R15	0.81	water	60	200	0	5	50	Marcotullio and de Jong (2010)
R16	0.99	water	8	30	61	64	280	Tamura et al. (2013)
R19	0.94	2-MTHF	50	220	1	2	7	Geilen et al. (2011)
R25	0.95	-	-	120	4	9	30	Yang et al. (2013)
R26	0.99	1,4-dioxane	-	140	2	4	22	Lanzafame et al. (2011)
R27	0.99	-	1	140	3	8	41	Fu et al. (2014)
R28	0.84	-	1	120	0	3	10	Liu et al. (2013)
R29	1.00	water	35	150	8	15	136	Dunesic et al. (2010)
R30	0.95	water	-	100	1	4	34	Badarinarayana et al. (2014)
R31	0.68	water/GBL	40	130	0	0	9	Cui et al. (2015)
R32	0.60	water	-	170	0	0	39	Ren et al. (2015)
R33	0.55	water/GBL	40	150	0	0	5	Cui et al. (2015)
R34	0.36	-	100	190	0	0	1	Al-Shaal et al. (2014)
R35	0.98	water	-	35	0	0	9	Ji et al. (2011)
R36	0.23	-	76	150	0	0	7	Manzer (2010)
R37	1.00	DCM	-	25	9	38	246	Noureldin and Lee (1982)
R38	0.75	water	-	50	0	0	0	Smith and Liao (2011)
R39	0.76	DCM	-	25	0	1	10	Thilagavathi and Jayabalakrishnan (2010)
R40	0.72	water	195	195	0	1	22	Potvin et al. (2011)
R41	0.92	water	1	250	1	1	62	Zhang et al. (2012)
R42	0.68	water	-	300	0	0	3	Gong et al. (2010)

C.3 Energy demand of separations

Energy requirement for all thermal separations are given in Table C.3 normalized on a feed rate of 1 kmol/s. The molar compositions (x) are given for all components (Comp.) as well as type and pressure of separation. For pressure-swing distillation (PSD) for both columns the respective pressure levels are presented, while for heteroazeotropic distillation (HD) both columns operate at the same pressure level. Flash evaporation (F) is used for the separation of fructose and water. In case an efficient vapor recompression (VRC) is feasible, energy demands are given in Table C.4.
As the transition from reaction R8 to R9, from reactions R30, R32, R40 to R29, from R35 to R42 and from R11 to R30 and R31 only require an increase in the concentration and not a complete water removal, additional separation fluxes are implemented for the adaption of the concentration. The respective energy demands are presented in Table C.5. As the concentration of reaction R9 is anyway higher than of the subsequent reaction R29, no separation is required in this case. For the sake of brevity, the following chemicals are abbreviated: 2,5-hydroxymethylfurfural (HMF), dimethylsulfoxid (DMSO), 2-methyltetrahydrofuran (2-MTHF), ethyllevulinate (EL), levulinic acid (LA), γ-valerolactone (GVL), γ-butyrolactone (GBL) and methylenechlorid (DCM).

C Case study parameters

Table C.3: Non-heat integrated energy requirement of all thermal separations.

Sep.	Reaction	Comp. 1	Comp. 2	Comp. 3	Comp. 4	x1	x2	x3	x4	Type	p [atm]	Heating [MW]	Cooling [MW]	Electricity [MW]
1	R12	HMF	DMSO	water	-	0.071	0.715	0.214	-	D	1	48	46	0
2	R12	water	DMSO	-	-	0.231	0.769	-	-	D	1	20	15	0
3	R8	HMF	water	-	-	0.040	0.960	-	-	D	1	40	40	0
4	R19	furfurylalcohol	2-MTHF	-	-	0.024	0.976	-	-	D	1	2	1	0
5	R16	furfurylalcohol	water	-	-	0.018	0.982	-	-	D	1	133	132	0
6	R10	EL	water	ethanol	-	0.333	0.334	0.333	-	D	1	38	28	0
7	R10	ethanol	water	-	-	0.500	0.500	-	-	PSD	1/10	690	690	0
8	R25	EL	ethanol	-	-	0.029	0.971	-	-	D	1	39	39	0
9	R26	EL	ethanol	-	-	0.042	0.958	-	-	D	1	39	38	0
10	R28	EL	formic acid	water	ethanol	0.004	0.004	0.008	0.984	D	1	51	51	0
11	R28	ethanol	water	ethanol	-	0.004	0.008	0.988	-	D	5	107	107	0
12	R28	LA	water	-	-	0.992	0.008	-	-	D	1/10	217	217	0
13	R9	LA	formic acid	water	-	0.048	0.048	0.904	-	D	1	40	38	0
14	R9	LA	water	-	-	0.050	0.950	-	-	D	9	234	234	0
15	R30	formic acid	formic acid	water	-	0.017	0.017	0.966	-	D	1	41	40	0
16	R30	LA	water	-	-	0.018	0.982	-	-	D	9	240	240	0
17	R32	formic acid	formic acid	water	-	0.003	0.003	0.994	-	D	1	41	41	0
18	R32	LA	water	water	-	0.003	0.997	-	-	D	9	240	240	0
19	R40	LA	formic acid	water	-	0.007	0.007	0.986	-	D	1	41	41	0
20	R40	formic acid	water	-	-	0.007	0.993	-	-	D	9	241	241	0
21	R27	GVL	ethanol	dioxane	-	0.077	0.077	0.846	-	D	2	36	34	0
22	R27	ethanol	dioxane	-	-	0.083	0.917	-	-	D	1	22	22	0
23	R29	GVL	water	-	-	0.028	0.972	-	-	D	1	97	96	0
24	R31.R33	GVL	GBL	water	-	0.024	0.452	0.524	-	D	1	36	29	0
25	R31.R33	GVL	GBL	-	-	0.050	0.950	-	-	D	1	894	894	0
26	R34.R36	2-butanol	water	-	-	0.500	0.500	-	-	HD	0.1	125	123	0
27	R38	iso-butanol	water	-	-	0.007	0.993	-	-	HD	1	6	5	0
28	R37	2-butanone	DCM	-	-	0.016	0.984	-	-	D	1	30	30	0
29	R39	2-butanone	DCM	-	-	0.003	0.997	-	-	D	1	30	30	0
30	R41	2-butanone	water	-	-	0.167	0.833	-	-	HD	3	40	39	0
31	R42	2,3-butanediol	water	-	-	0.003	0.997	-	-	HD	0.5	7	5	0
32	R35	fructose	water	-	-	0.019	0.981	-	-	D	1	50	49	0
33	R11		water	-	-	0.002	0.998	-	-	F	1	48	0	0

C.3 Energy demand of separations

Table C.4: Energy requirement for heat-integrated thermal separation by means of VRC. Only separations for which VRC is an efficient technique for heat integration are presented.

Sep.	Reaction	Comp. 1	Comp. 2	Comp. 3	Comp. 4	x1	x2	x3	x4	Type	p [atm]	Heating [MW]	Cooling [MW]	Electricity [MW]
34	R4, R5	water	ethanol	-	-	0.978	0.022	-	-	PSD	1/10	1	0	2
35	R12	water	DMSO	-	-	0.231	0.769	-	-	D	1	5	0	5
36	R8	HMF	water	-	-	0.040	0.960	-	-	D	1	1	0	14
37	R16	furfurylalcohol	water	-	-	0.018	0.982	-	-	D	1	0	0	36
38	R10	ethanol	water	-	-	0.500	0.500	-	-	PSD	1/10	4	0	52
39	R25	EL	ethanol	-	-	0.029	0.971	-	-	D	1	1	0	18
40	R26	EL	ethanol	-	-	0.042	0.958	-	-	D	1	1	0	18
41	R28	EL	formic acid	water	ethanol	0.004	0.004	0.008	0.984	D	1	0	0	24
42	R28	formic acid	water	ethanol	-	0.004	0.008	0.988	-	D	5	0	0	15
43	R28	ethanol	water	-	-	0.992	0.008	-	-	D	1/10	19	0	4
44	R9	formic acid	water	-	-	0.050	0.950	-	-	D	9	0	0	15
45	R30	formic acid	water	-	-	0.018	0.982	-	-	D	9	0	0	15
46	R32	formic acid	water	-	-	0.003	0.997	-	-	D	9	0	0	15
47	R40	formic acid	water	-	-	0.007	0.993	-	-	D	9	0	0	15
48	R27	ethanol	dioxane	-	-	0.083	0.917	-	-	D	2	0	0	3
49	R29	GVL	water	-	-	0.028	0.972	-	-	D	1	1	0	41
50	R31,R33	GVL	GBL	water	-	0.024	0.452	0.524	-	D	1	7	0	12
51	R31,R33	GVL	GBL	-	-	0.050	0.950	-	-	D	1	0	0	22
52	R37	2-butanone	DCM	-	-	0.016	0.984	-	-	D	1	0	0	5
53	R39	2-butanone	DCM	-	-	0.003	0.997	-	-	D	1	0	0	5
54	R35	2,3-butanediol	water	-	-	0.019	0.981	-	-	D	1	1	0	15
55	R11	fructose	water	-	-	0.002	0.998	-	-	E	1/0.3	22	0	2

C Case study parameters

Table C.5: Energy requirement for concentration increase is presented associated with the respective reactions.

Sep.	Reaction	Target R.	Comp. 1	Comp. 2	Comp. 3	x1	x2	x3	Type	p [atm]	Heating [MW]	Cooling [MW]	Electricity [MW]
56	R8	R9	HMF	water	-	0.040	0.960	-	D	1	6	6	0
57						0.045	0.955	-	VRC-D	1	0	0	0.1
58	R30	R29	LA	formic acid	water	0.045	0.955	0.966	D	1	611	611	0
59						0.017	0.017	0.944	VRC-D	1	0	0	11.2
60	R32	R29	LA	formic acid	water	0.028	0.028	0.944	D	1	108	108	0
61						0.028	0.028	0.944	VRC-D	1	0	0	15.5
62	R40	R29	LA	formic acid	water	0.003	0.003	0.994	D	1	251	251	0
63						0.028	0.028	0.944	VRC-D	1	0	0	31.5
64	R35	R41	2,3-butanediol			0.007	0.007	0.986	D	1	40	40	0
						0.028	0.028	0.944					
65	R11	R30	fructose	water	-	0.019	0.981	-	E	1	32	32	0
66						0.200	0.800	-	VRC-E	1	0	0	0.8
						0.002	0.998	-					
67	R11	R31	fructose	water	-	0.018	0.982	-	E	1	36	36	0
68						0.018	0.982	-	VRC-E	1	0	0	0.9
						0.002	0.998	-					
						0.048	0.952	-					
						0.048	0.952	-					

C.4 Properties

Table C.6 summarizes the pure-component properties required for the calculation of the binary ratios in the feasibility analysis. If not otherwise stated the boiling points T_B and vapor pressures p_{oi} (298 K) are taken from the Aspen database version 7.3., the octanol-water partition coefficient $\log K_{ow}$ from the Gestis database (Institute for Occupational Safety and Health of the German Social Accident Insurance, 2016). The exergy of a component,

$$E_{ex} = \Delta_f G^0 + N_C \cdot ex_C^0 + N_H \cdot ex_{H2}^0 + N_O \cdot ex_{O2}^0. \tag{C.1}$$

is calculated based on the Gibbs free energy of a molecule $\Delta_f G^0$, the number of carbon (N_C), hydrogen (N_H) and oxygen (N_O) atoms, which are multiplied with the molar standard exergy of the elements ex_x^0. The molar standard exergy of carbon, hydrogen and oxygen are 406 kJ/mol, 235 kJ/mol and 5 kJ/mol, respectively (Diederichsen, 1991). The product's $\Delta_f G^0$ are taken from Dean and Lange (1999). If the data is not included in any database, the data is estimated using ICAS version 15 (Gani et al., 1997), which is also utilized to determine the molar volume (298 K) and Hildebrandt solubility. Estimated NRTL parameters are presented in Table C.7.

Table C.6: Pure component properties.

component	T_B [K]	p_{oi} [atm]	$\log K_{ow}$ [-]	V_M [$\frac{cm^3}{mol}$]	HS [$MPa^{0.5}$]	$\Delta_f G^0$ [$\frac{kJ}{mol}$]	E_{ex} [$\frac{kJ}{mol}$]
water	373	0.0313	-	18	48		
ethanol	351	0.0782	-0.3	61	22	-175	1343
dimethylsulfoxid	464	0.0008	-1.35	78	27		
2,5-hydroxymethylfurfural	499	0.0003	0.02[1]	102	28[1]	-289	2855
furfurylalcohol	443	0.0008	0.28	87	25	-154	2583
2-methyltetrahydrofuran	353	0.1259	1.25	98	18		
ethyllevulinate	479	0.0002	0.67[1]	144	20	-436[1]	3818
formic acid	347	0.0561	-0.54	37	25	-361	284
levulinic acid	530	0.000003	-0.28[1]	108	24	-478[1]	2496
γ-valerolactone	481	0.0005	-0.133	95	21	-266[1]	2705
γ-butyrolactone	405	0.0006	-0.64	79	24		
1-4-dioxane	374	0.0503	-0.27	83	21		
methylenechlorid	313	0.5765	1.25	66	20		
2-butanol	373	0.0231	0.65	93	21	-177	2621
isobutanol	381	0.0138	0.76	95	21	-158[1]	2640
2-butanone	353	0.1215	0.29	91	19	-151	2412
2,3-butanediol	481	0.00005	-0.92	90	25	-294[1]	2506
fructose	647	0.00002	-3.52[1]	114	38	-844	3012

[1]The value has been estimated using ICAS version 15 Gani et al. (1997).

C Case study parameters

Table C.7: Estimated NRTL parameter using COSMO-RS (Klamt, 2005).

Component 1	Component 2	alpha	a12	a21	b12	b21
2,5-hydroxymethylfurfural	dimethylsulfoxid	0.3	1.40	-3.11	-1684.16	2839.41
water	2,5-hydroxymethylfurfural	0.3	8.21	-1.00	-1878.60	127.76
furfurylalcohol	2-methyltetrahydrofuran	0.3	-0.74	3.52	-118.88	-780.53
ethyllevulinate	water	0.3	-0.96	-4.15	391.89	-28.80
ethyllevulinate	ethanol	0.3	-0.39	-0.85	33.64	680.74
ethyllevulinate	formic acid	0.3	-0.14	2.85	-606.53	-775.86
water [1]	formic acid [1]	0.3	0.00	0.00	323.99	-220.50
ethanol	formic acid	0.3	13.88	-19.83	-5449.94	7236.56
levulinic acid	levulinic acid	0.3	-0.61	1.63	-413.11	-306.30
water	γ-valerolactone	0.3	-1.72	-0.39	1634.18	-114.06
ethanol	1-4-dioxane	0.3	0.72	-1.29	99.51	377.14
γ-valerolactone	γ-valerolactone	0.3	-4.31	2.41	2372.32	-1351.78
water	γ-valerolactone	0.3	3.85	-1.11	-370.78	379.91
γ-butyrolactone	γ-valerolactone	0.3	0.25	-0.22	0.01	0.01
fructose	water	0.3	-1.30	8.71	-101.51	-2752.23

[1] NRTL parameter have been estimated using UNIFAC.

C.5 Active pathway fluxes

In the following, the active fluxes obtained using RNFA are given in Table C.8 at the specific points of each Pareto curve. The active fluxes obtained using the PNFA for the scenario I) without residue combustion are shown in Table C.9. The active fluxes for the scenario II), which includes residue combustion, are given in Table C.10. The point of minimal cost is referred to as *min Cost*, the point of minimal EI or CED are marked accordingly. The points of the Pareto curve are numbered, starting with the point of minimal cost (1) and ending with the point (6) of either minimal EI (RNFA) or minimal CED (PNFA). For clarity reason, reactions R1 - R3 are not explicitly stated, but are active for all processes at all points of the Pareto curve in both analysis (RNFA, PNFA). For the same reason residue combustion is not explicitly listed, but is activated in scenario II) for all points of the Pareto curve with the only exception of point (1) of the ethyllevulinate process.

Table C.8: Active reaction fluxes in the RNFA for all points of the Pareto curve.

product	min cost (1)	2	3	4	5	min EI (6)
ethanol	5,6	4-7	4-7	4-7	4-7	4-7
iso-butanol	5,38	5,38	5,38	5,38	5,38	5,38
2-butanone	5,35,41	5,35,41	5,35,41	5,35,41	5,35,41	5,35,41
EL	5,6,11,12,26	4-7,11,12,26	4-7,11,12,26	4-8,11,12,26	4-7,9-12,26,40	4-12,15,16,25,28,30,32,40
GVL	29,40	5,11,29,30,40	5,11,29,30,40	5,11,29,30,40	5,9,11,12,29,40	5,8,9,11,12,29-33,40
2-butanol	29,34,40	5,9,11,12,29,34,40	5,9,11,12,29,34,40	5,9,11,12,29,34,40	5,9,11,12,29,34,40	5,8,9,11,12,29-34,40

Table C.9: Active reaction and separation fluxes in the PNFA for all points of the Pareto curve. Residue combustion for an internal energy supply is not considered.

product	fluxes	min cost	2	3	4	5	min CED
ethanol	reaction	4-7	4-7	4-7	4-7	4-7	4-7
ethanol	separation	34	34	34	34	34	34
iso-butanol	reaction	5,38	5,38	5,38	5,38	5,38	5,38
iso-butanol	separation	27	27	27	27	27	27
2-butanone	reaction	5,35,41	5,35,41	5,35,41	5,35,41	5,35,41	5,35,41
2-butanone	separation	30,64	30,54	30,54	30,54	30,54	30,54
EL	reaction	4,5,7,8,26	4,5,7,8,26	4,5,7,8,26	4,5,7,8,26	4,5,7-10	4,5,7-10
EL	separation	3,9,34	3,9,34,36	34,36,40	6,14,34,38,56	6,34,38,56	6,34,38,57
GVL	reaction	5,33	5,33	5,33	5,33	5,33	5,33
GVL	separation	24,51	24,50,51	24,50,51	24,50,51	24,50,51	50,51
2-butanol	reaction	5,33,34	5,33,34	5,33,34	5,33,34	5,33,34	5,33,34
2-butanol	separation	24,51	24,50,51	24,50,51	24,50,51	24,50,51	50,51

C.5 Active pathway fluxes

Table C.10: Active reaction and separation fluxes in the PNFA for all points of the Pareto curve. Residue combustion for an internal energy supply is taken into account.

product	fluxes	min cost	2	3	4	5	min CED
ethanol	reaction	5-6	5-6	5-6	5-6	5-6	4-7
	separation	34	34	34	34	34	34
iso-butanol	reaction	5,38	5,38	5,38	5,38	5,38	5,38
	separation	27	27	27	27	27	27
2-butanone	reaction	5,35,41	5,35,41	5,35,41	5,35,41	5,35,41	5,35,41
	separation	30,54	30,54	30,54	30,54	30,54	30,54
EL	reaction	5,6,8,26	5,6,8,26	5,6,8,26	5,6,8,26	5,6,8,26	5,6,8,26
	separation	3,9,34	3,9,34	3,9,34	3,9,34	3,9,34	9,34,36
GVL	reaction	5,33	5,33	5,33	5,33	5,33	5,33
	separation	24,51	24,51	24,51	24,51	24,51	50,51
2-butanol	reaction	5,8,9,29,33,34	5,8,9,29,33,34	5,8,9,29,33,34	5,8,9,29,33,34	5,33,34	5,33,34
	separation	24,49,51,56	24,49,51,56	24,49,51,56	24,49,51,56	24,51	50,51

C Case study parameters

C.6 Supply chain data and design

In the following, the procedure to determine the biomass availability per area is briefly described. The procedure is adapted from Schwaderer (2012) and visualized in Figure C.1.

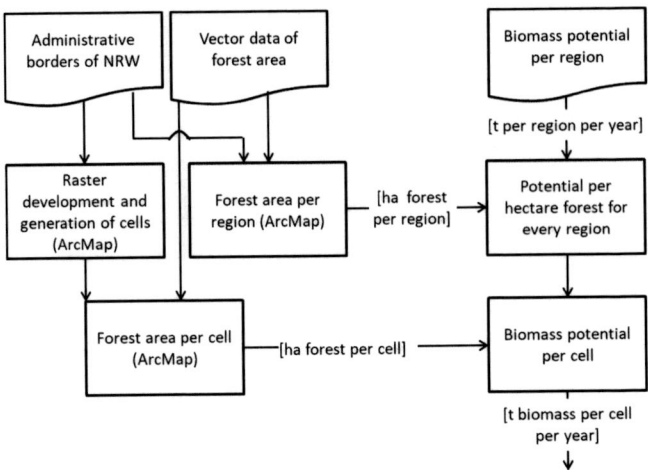

Figure C.1: Data extraction and processing to obtain the biomass availability and distribution for NRW, which is adapted from Schwaderer (2012).

Here, the data of Dieter et al. (2001) is utilitzed for quantifying the biomass potential. Dieter et al. (2001) estimate an overall residual wood potential of 1.5 million dry tons per year in NRW. In addition, data for the different administrative regions in NRW is provided. This data further needs to be combined with information on the administrative borders (Bundesamt für Kartographie und Geodäsie, 2016) and the distribution of the forest area (Deutsches Zentum für Luft- und Raumfahrt, 2016) assuming a residual wood allocation similar to the forest area distribution. Finally, the geoinformation system ArcMap version 10.4 (Esri, 2016) is utilized to create a grid for the different cell sizes and to determine the forest area per region and cell. In combination with the data of Dieter et al. (2001) the biomass potential per region and, thus, per cell is obtained.

In Table C.11 a literature overview on fixed and variable prices for truck transportation in dependency of the region, year and biomass type is given. Table C.12 presents all chemical sites, which are analyzed for a potential biorefinery. The sites are given along with the addresses and Gauss-Krueger coordinates.

C.6 Supply chain data and design

Table C.11: Price parameters for truck transportation available in literature.

Country	$P_{BT,fix}$ [$\frac{\$}{t}$]	$P_{BT,var}$ [$\frac{\$}{t\,km}$]	type	w_{H2O} [wt%]	Reference
Germany	2.69	0.15	residual wood	35-50	Schwaderer (2012)
Sweden	2.05	0.09	logging residues	NA	Angus-Hankin et al. (1995)
Sweden	7.28	0.16	logging residues	0	Börjesson (1996), Börjesson and Gustavsson (1996)
Netherlands	1.63	0.02	logs,bales,chips	NA	Hamelinck et al. (2005)
Canada	3.01	0.07	wood chips	45	Searcy et al. (2007)
Canada	4.98	0.11	wood chips	0	Mahmudi and Flynn (2006)
USA	4.84	0.46	biomass	NA	Garcia and You (2015a)
USA	-	0.16	logging residues	NA	Kim et al. (2011a)
USA	3.01	0.07	wood chips	45	Marvin et al. (2013a)
USA	10.7	0.19	corn stover	0	Ng et al. (2016)
USA	5.7	0.14	corn stover	NA	Ileleji et al. (2010)
USA	-	0.125	woody biomass	NA	Ekşioğlu et al. (2009)
Brazil	-	0.36	sugarcane	NA	Jonker et al. (2016)

Table C.12: Location of production sites

Number	City	Street	X Coordinates	Y Coordinates
1	Heinsberg	Boos-Fremery-Straße	3299742.85	5665138.64
2	Hürth	Goldenbergstraße	3348188.84	5638381.52
3	Marl	Paul-Baumann-Straße	3368631.59	5729030.43
4	Leverkusen	Kaiser Wilhelm Allee	3358719.04	5655012.03
5	Dormagen	Parallelweg	3348917.89	5660893.45
6	Krefeld	Friedensstraße	3336160.66	5695281.65
7	Niederkassel	Feldmühlestraße	3361141.8	5633978.32
8	Düsseldorf	Henkelstraße	3349571.62	5673205.41
9	Köln	Emdener Straße	3356200.17	5655930.54
10	Rheinberg	Xantener Straße	3332071.55	5717112.32

The following table presents for all cells, the coordinates based on a grid length of 20 km. To determine the coordinates, the middle point of each cell has been used. If no street is available in the center point of a cell, the coordinates are relocated to the closest street available. These relocation coordinates are given in the table as well. Furthermore, the annual biomass availability per grid is given. The data is based on the assumption of biomass as a homogeneous good. To keep problem size at a reasonable level, it is further assumed that harvesting cost are constant and equal for all delivered biomass, thereby neglecting that typically easily accessible resources show lower harvesting cost, and might, thus, be exploited first. Thus, a common biomass market price is utilized independent of the grid. The table shows as well active cells for an ethanol production process excluding combustion (process I) and including combustion (process II).

C Case study parameters

Table C.13: Biomass availability per location with a grid length of 20 x 20 km

ID	Center coordinates		Relocation coordinates		Biomass $[\frac{t}{year}]$	Process I $[\frac{t}{year}]$	Process II $[\frac{t}{year}]$
	X	Y	X	Y			
0	3310359	5589507	3311295	5590301	12676	0	0
1	3330359	5589507	3330357	5589607	24298	0	0
2	3350359	5589507	3349117	5592153	6437	0	0
3	3290359	5609507	3299489	5606456	393	0	0
4	3310359	5609507	3310360	5609515	32810	0	0
5	3330359	5609507	3330363	5609501	23228	0	0
6	3350359	5609507	3350368	5609504	21066	0	0
7	3370359	5609507	3370354	5609487	10402	10402	10402
8	3390359	5609507	3385909	5611104	1249	1249	1249
9	3430359	5609507	3432366	5616135	658	0	0
10	3290359	5629507	3290449	5629488	3883	0	0
11	3310359	5629507	3310366	5629519	27200	0	0
12	3330359	5629507	3330369	5629483	4106	4106	4106
13	3350359	5629507	3350372	5629412	8533	8533	8533
14	3370359	5629507	3370395	5629482	8552	8552	8552
15	3390359	5629507	3390310	5629507	18035	18035	18035
16	3410359	5629507	3408123	5630313	6081	0	0
17	3430359	5629507	3430386	5629502	30634	0	0
18	3450359	5629507	3450629	5634852	8537	0	0
19	3290359	5649507	3290383	5649531	2758	0	0
20	3310359	5649507	3310382	5649453	3231	0	0
21	3330359	5649507	3330351	5649496	6900	6900	6900
22	3350359	5649507	3350360	5649507	5688	5688	5688
23	3370359	5649507	3370334	5649525	25958	25958	25958
24	3390359	5649507	3390326	5649504	20117	20117	20117
25	3410359	5649507	3410348	5649489	29392	4176	16752
26	3430359	5649507	3430342	5649497	46216	0	0
27	3450359	5649507	3450340	5649576	51772	0	0
28	3470359	5649507	3467997	5652349	7972	0	0
29	3290359	5669507	3291394	5666524	3705	0	0
30	3310359	5669507	3310371	5669553	10963	0	0
31	3330359	5669507	3330387	5669510	3674	3674	3674
32	3350359	5669507	3350278	5669538	9077	9077	9077
33	3370359	5669507	3370348	5669516	23797	23797	23797
34	3390359	5669507	3390326	5669393	25438	25438	25438
35	3410359	5669507	3410427	5669541	42575	0	0
36	3430359	5669507	3430352	5669504	45411	0	0
37	3450359	5669507	3450358	5669508	48719	0	0
38	3470359	5669507	3470363	5669511	40814	0	0
39	3490359	5669507	3484969	5671381	1208	0	0
40	3290359	5689507	3294769	5683227	4350	0	0
41	3310359	5689507	3310354	5689484	10808	0	0
42	3330359	5689507	3330396	5689522	3851	3851	3851
43	3350359	5689507	3350348	5689589	13761	13761	13761
44	3370359	5689507	3370421	5689497	12661	12661	12661
45	3390359	5689507	3390375	5689445	31338	31338	31338
46	3410359	5689507	3410338	5689561	44936	0	0
47	3430359	5689507	3430381	5689523	45716	0	0
48	3450359	5689507	3450387	5689519	47936	0	0
49	3470359	5689507	3470345	5689502	32692	0	0
50	3490359	5689507	3489525	5692549	5527	0	0
51	3310359	5709507	3310291	5709517	4346	0	0
52	3330359	5709507	3330385	5709486	6304	0	0
53	3350359	5709507	3350371	5709517	6605	6605	6605
54	3370359	5709507	3370348	5709517	4746	0	0
55	3390359	5709507	3390371	5709530	5011	0	0
56	3410359	5709507	3410354	5709598	6932	0	0

C.6 Supply chain data and design

ID	Center coordinates		Relocation coordinates		Biomass	Process I	Process II
	X	Y	X	Y	$[\frac{t}{year}]$	$[\frac{t}{year}]$	$[\frac{t}{year}]$
57	3430359	5709507	3430316	5709504	20632	0	0
58	3450359	5709507	3450546	5709444	18697	0	0
59	3470359	5709507	3470330	5709551	33675	0	0
60	3490359	5709507	3490459	5709570	32581	0	0
61	3510359	5709507	3510357	5709507	6473	0	0
62	3530359	5709507	3525657	5712990	1012	0	0
63	3290359	5729507	3292252	5730224	5273	0	0
64	3310359	5729507	3310470	5729742	8973	0	0
65	3330359	5729507	3330381	5729509	5358	0	0
66	3350359	5729507	3350377	5729503	17849	0	0
67	3370359	5729507	3370386	5729509	20046	0	0
68	3390359	5729507	3390411	5729500	9281	0	0
69	3410359	5729507	3410536	5729527	3211	0	0
70	3430359	5729507	3430382	5729325	6329	0	0
71	3450359	5729507	3450323	5729769	2000	0	0
72	3470359	5729507	3470407	5729493	4845	0	0
73	3490359	5729507	3490348	5729590	22716	0	0
74	3510359	5729507	3510292	5729536	244851	0	0
75	3530359	5729507	3529029	5730410	8164	0	0
76	3290359	5749507	3290469	5749227	3105	0	0
77	3310359	5749507	3310379	5749494	1732	0	0
78	3330359	5749507	3330446	5749461	2811	0	0
79	3350359	5749507	3350354	5749521	8812	0	0
80	3370359	5749507	3370249	5749541	13049	0	0
81	3390359	5749507	3390309	5749486	10253	0	0
82	3410359	5749507	3410379	5749534	12658	0	0
83	3430359	5749507	3430443	5749511	7378	0	0
84	3450359	5749507	3450361	5749577	4385	0	0
85	3470359	5749507	3470363	5749509	12383	0	0
86	3490359	5749507	3490346	5749433	29509	0	0
87	3510359	5749507	3510357	5749482	17537	0	0
88	3530359	5749507	3530462	5749698	7908	0	0
89	3350359	5769507	3350358	5769486	7354	0	0
90	3370359	5769507	3370301	5769485	10117	0	0
91	3390359	5769507	3390272	5769858	7092	0	0
92	3410359	5769507	3410377	5769515	8554	0	0
93	3430359	5769507	3430042	5769193	6099	0	0
94	3450359	5769507	3450347	5769491	6268	0	0
95	3470359	5769507	3470354	5769507	6211	0	0
96	3490359	5769507	3490325	5769619	9881	0	0
97	3510359	5769507	3510354	5769534	9232	0	0
98	3350359	5789507	3354656	5783423	432	0	0
99	3370359	5789507	3370355	5789507	4282	0	0
100	3390359	5789507	3390334	5789478	8723	0	0
101	3410359	5789507	3410328	5789467	10547	0	0
102	3430359	5789507	3427866	5790231	6871	0	0
103	3470359	5789507	3470368	5789478	6468	0	0
104	3490359	5789507	3490367	5789491	6450	0	0
105	3510359	5789507	3506336	5788799	846	0	0
106	3390359	5809507	3391512	5805416	492	0	0
107	3410359	5809507	3410381	5809554	2308	0	0
108	3430359	5809507	3427734	5806065	2775	0	0
109	3450359	5809507	3452244	5809002	483	0	0
110	3470359	5809507	347032	5809516	3993	0	0
111	34903592	5809507	3489959	5809190	3086	0	0
112	3510359	5809507	3510129	5810339	767	0	0
113	3470359	5829507	3471983	5822600	0	0	0

C.7 Multi-product biorefinery

For the assessment of a multi-product biorefinery with a sale of value-added chemicals, the original data on market sizes and prices for all traded chemicals are given in Table C.15. The prices are updated using the Producer Price Index for chemicals (cf. Table C.1). The monthly values thereof are averaged over a time period of one year. The market data is updated using the component specific CAGRs. Therefore, references publishing both, the market size and associated CAGR are preferred. The resulting market sizes and prices for the reference year are listed in Table C.16. For γ-valerolactone price and market data is not available in the open literature.

Table C.15: Prices and market sizes ms of chemicals.

	Price $[\frac{\$}{t}]$	ms $[\frac{t}{year}]$	CAGR [%]	Year	Reference
ethyllevulinate	12500			2016	Alibaba (2016)
		32	5.3	2014	Grand View Research (2016e)
furfural	1000			2006	Hayes et al. (2006)
		300,000	11.9	2013	Grand View Research (2016a)
furfuryl alcohol	1760			2006	Hayes et al. (2006)
		258,900	11.9	2013	Grand View Research (2016a)
isobutanol	2359			2014	ICIS (2016)
		552,400	6.0	2014	Grand View Research (2016b)
2-butanone	3316			2014	S and P Global (2016)
		1,420,000	4.5	2014	Grand View Research (2016d)
levulinic acid	5000			2006	Hayes et al. (2006)
		2606	5.7	2013	Grand View Research (2016c)
formic acid	1350			2006	Hayes et al. (2006)
		1,733,000	5.3	2017	Front Research (2016)
2,3-butanediol	15,788	61,800	3,2	2012	Transparency Market Research (2016)

Table C.16: Prices and market sizes ms of chemicals for the reference year 2014.

	Price $[\frac{\$}{t}]$	ms $[\frac{t}{year}]$
2,5-hydroxymethylfurfural	3038	96
levulinic acid	6804	2,755
formic acid	1506	1,484,272
ethyllevulinate	15888	32
furfural	1361	335,700
furfuryl alcohol	2395	259,208
2-butanol	1907	1,375,000
2,3-butanediol	14863	65,818
butanone	1176	137,5000
iso-butanol	2359	552,400

C.8 Mixtures

In the following table, the final molar mixture compositions of the two co-production scenarios are presented.

Table C.17: Molar compositions of the resulting mixtures derived in the co-production scenario analyses.

	scenario I		scenario II	
	min cost SI	min CED CI	min cost CI	min CED CI
acetone	0.244	-	-	-
1-butanol	0.701	-	-	-
ethanol	0.049	-	-	-
water	0.006	-	0.054	-
DNBE	-	-	0.677	-
octane	-	0.010	-	0.010
octanol	-	0.250	0.133	0.250
BTHF	-	0.010	0.015	0.010
DOE	-	0.730	0.121	0.730
Sum	1	1	1	1

C.9 Fermentation data

The fermentation products are ranked and analyzed based on their process energy demand (CED) and the specific production costs. Figure C.2 summarizes potential separation steps long with an internal recycle structure for purifying diluted fermentation broths. The specific production cost TAC_{spec} include annualized investment, raw material, auxiliary, utility and waste disposal cost, which are summed up and normalized on the heating value of the products, according to:

$$TAC_{spec} = \frac{TAC}{b_{product}\Delta h_{comb,product}}. \tag{C.2}$$

In Table C.18 all analyzed fermentation products and the data-set with the highest yield, productivity and titer available in literature, are presented. The respective entries are marked in bold to visualize the highest value achieved so far. In addition, the data-set responsible for the lowest energy demand for fermentation and downstream processing is marked in bold as well. HPA and malic acid are exceptions, as the data-set is similar for all three performance indicators (yield, productivity, titer). In Table C.19 and C.20 all data-sets available in literature for the production of itaconic and succinic acid are summarized, respectively. The respective data-set responsible for the

C Case study parameters

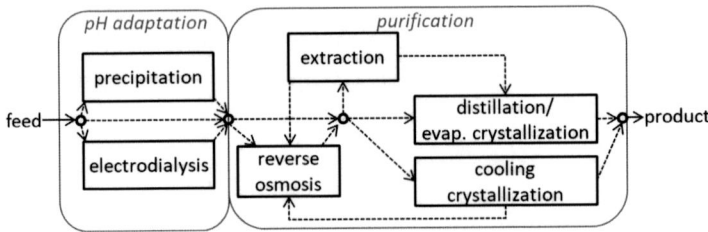

Figure C.2: Overview on potential downstream processing steps for a purification of fermentation products.

lowest energy demand is again marked in bold. All fermentations utilize glucose, if not stated otherwise. Table C.21 then presents the required property data of all analyzed components. The logarithm of the acid dissociation constant pKs is mandatory to calculate the degree of dissociation.

Although predictive solubility models exist in literature, regression of experimental data is applied in this work. According to Atkins and de Paula (2014), the solubility can be determined based on the heat of fusion and the melting point. Another approach is the general solubility equation of Jain et al. (2006), which is based on a component's $\log K_{ow}$ value and melting point. However, for the analyzed carboxylic acids, the models either under- or overestimate the solubilities significantly, as shown in Figure C.3. Therefore, experimental values have been gathered and compared. The specific references are listed in Table C.22 and the utilized experimental data for regression is marked in bold. For regression, an exponential approach is deemed most suitable. The exponential approach determines the temperature-dependent solubility in water Sol, based on a pre-factor and an exponential factor, according to:

$$Sol = Sol_1 \cdot \exp(Sol_2 T). \tag{C.3}$$

The lower ($T_{Sol,low}$) and upper temperature limit $T_{Sol,up}$ of the experimental solubility data are given in Table C.21.

Table C.18: Data-sets of all analyzed fermentation products. For each product, a data-set reporting the highest yield, productivity and titer are included in the analysis. The highest values are marked bold. Furthermore, the data-set of the lowest overall energy requirement is marked bold as well.

product	yield [g/g]	productivity [g/L/h]	titer [g/L]	pH	aeration	T [°C]	microorganism	reference
ethanol	0.5	1.8	170	NA	anaerobic	20	*Saccharomyces cerevisiae*	Thomas and Ingledew (1992)
ethanol	0.5	**82**	0.4	3.8	anaerobic	35	*Saccharomyces cerevisiae*	Cysewski and Wilke (1977)
ethanol	0.5	1.8	**170**	NA	anaerobic	20	*Saccharomyces cerevisiae*	Thomas and Ingledew (1992)
2,3-butanediol	0.5	0.3	86	5.0	anaerobic	30	*Klebsiella oxytoca*	Qureshi and Cheryan (1989)
2,3-butanediol	0.49	5.4	110	6.3	anaerobic	35	*Enterobacter aerogenes*	Zeng et al. (1991)
2,3-butanediol	0.48	4.2	150	6.0	anaerobic	37	*Klebsiella pneumoniae*	Ma et al. (2009)
1,3-propanediol[1]	**0.56**	1.2	65	6.5	anaerobic	35	*Clostridium butyricum*	Saint-Amans et al. (1994)
1,3-propanediol	0.51	**3.5**	135	NA	anaerobic	35	not disclosed	Nakamura and Whited (2003)
1,3-propanediol	0.51	3.5	**135**	NA	anaerobic	35	not disclosed	Nakamura and Whited (2003)
itaconic acid	**0.62**	0.5	86	1.7	aerobic	33	*Aspergillus terreus*	Kuenz et al. (2012)
itaconic acid	0.58	**1.2**	129	3.0	aerobic	33	*Aspergillus terreus*	Hevekerl et al. (2014)
itaconic acid	0.59	0.5	**146**	3.0	aerobic	33	*Aspergillus terreus*	Hevekerl et al. (2014)
succinic acid	**1.10**	1.3	99	7.0	anaerobic	37	*Escherichia coli*	Vemuri et al. (2002)
succinic acid	0.89	**10.4**	83	6.4	anaerobic	39	*Anaerobiospirillum succiniproducens*	Meynial-Salles et al. (2008)
succinic acid	0.92	3.2	**146**	6.0	anaerobic	35	*Corneybacterium glutamicum*	Okino et al. (2008)
fumaric acid	0.88	4.3	85	4.5	aerobic	20	*Rhizopus oryzae*	Cao et al. (1996)
fumaric acid	0.88	**4.3**	85	4.5	aerobic	35	*Rhizopus oryzae*	Cao et al. (1996)
fumaric acid	0.37	1.0	**121**	7.0	aerobic	20	*Rhizopus arrhizus*	Ling and Ng (1989)
lactic acid	**0.99**	3.5	183	7.5	anaerobic	37	*Bacillus subtilis*	Gao et al. (2012)
lactic acid	0.98	**160**	57	6.3	anaerobic	42	*Lactobacillus delbrueckii*	Ohleyer et al. (1985)
lactic acid	0.92	1.8	**231**	5.7	anaerobic	37	*Rhizopus oryzae*	Yamane and Tanaka (2013)
HPA[1]	**0.65**	1.7	49	6.9	anaerobic	37	*Klebsiella pneumoniae*	Huang et al. (2013)
HPA[1]	0.65	**1.7**	49	6.9	anaerobic	37	*Klebsiella pneumoniae*	Huang et al. (2013)
HPA[1]	0.65	1.7	**49**	6.9	anaerobic	37	*Klebsiella pneumoniae*	Huang et al. (2013)
malic acid	**1.03**	0.9	154	6.1	aerobic	20	*Saccharomyces cerevisiae*	Brown et al. (2013)
malic acid	1.03	**0.9**	154	6.1	aerobic	35	*Saccharomyces cerevisiae*	Brown et al. (2013)
malic acid	1.03	0.9	**154**	6.1	aerobic	20	*Saccharomyces cerevisiae*	Brown et al. (2013)

[1]Fermentation based on glycerol.

Table C.19: Summary of all literature-known data-sets for the production of itaconic acid.

number	yield [g/g]	productivity [g/L/h]	titer [g/L]	pH	aeration	T [°C]	microorganism	reference
1	0.58	1.15	129	3.0	aerobic	33	Aspergillus terreus	Hevekerl et al. (2014)
2	0.53	0.41	87	1.7	aerobic	33	Aspergillus terreus	Hevekerl et al. (2014)
3	0.59	0.48	146	3.0	aerobic	33	Aspergillus terreus	Hevekerl et al. (2014)
4	0.54	0.57	82	2.0	aerobic	37	Aspergillus terreus	Yahiro et al. (1995)
5	0.62	0.37	49	3.1	aerobic	37	Aspergillus terreus	Welter (2000)
6	0.58	0.29	90	1.7	aerobic	33	Aspergillus terreus	Kuenz et al. (2012)
7	0.55	0.32	30	2.0	aerobic	36	Aspergillus terreus	Kautola et al. (1985)
8	0.43	0.31	49	2.0	aerobic	40	Aspergillus terreus	Park et al. (1993)
9	0.62	0.51	86	1.7	aerobic	33	Aspergillus terreus	Kuenz et al. (2012)
10	0.45	0.26	44	2.0	aerobic	NA	Aspergillus terreus	Okabe et al. (1993)
11	0.49	0.47	56	2.0	aerobic	NA	Aspergillus terreus	Park et al. (1994)
12	0.44	0.31	51	2.0	aerobic	NA	Aspergillus terreus	Park et al. (1994)
13	0.39	0.39	46	2.0	aerobic	NA	Aspergillus terreus	Park et al. (1994)
14	0.48	0.37	48	2.0	aerobic	NA	Aspergillus terreus	Park et al. (1994)
15	0.18	1.20	12	2.0	aerobic	36	Aspergillus terreus	Kautola et al. (1985)
16[1]	0.23	0.12	14	2.0	aerobic	36	Aspergillus terreus	Kautola et al. (1985)
17	0.34	0.26	51	3.1	aerobic	37	Aspergillus terreus	Kautola et al. (1991)
18[1]	0.04	0.07	8	2.5	aerobic	36	Aspergillus terreus	Kautola (1990)
19	0.46	0.73	18	3.0	aerobic	36	Aspergillus terreus	Ju and Wang (1986)
20	0.18	0.46	10	6.8	aerobic	33	Ustilago maydis	Wewetzer (2014)
21	0.18	0.68	7	6.8	aerobic	33	Ustilago maydis	Wewetzer (2014)
22	0.17	0.27	20	6.8	aerobic	30	Ustilago maydis	Klement et al. (2012)
23[2]	0.36	0.16	54	3.5	aerobic	30	Aspergillus terreus	Elnaghy and Megalla (1975)
24[2]	0.12	0.09	16	3.0	aerobic	36	Aspergillus terreus	Kautola et al. (1990)

[1] Fermentation based on xylose.
[2] Fermentation based on sucrose.

C.9 Fermentation data

Table C.20: Summary of all literature-known data-sets for the production of succinic acid.

number	yield [g/g]	productivity [g/L/h]	titer [g/L]	pH	aeration	T [°C]	microorganism	reference
1	0.82	1.36	106	6.1	anaerobic	39	*Actinobacillus succinogenes*	Guettler and Jain (1996)
2	0.86	0.88	34	6.8	anaerobic	38	*Actinobacillus succinogenes*	Urbance et al. (2004)
3	0.50	0.30	4	6.4	anaerobic	NA	*Actinobacillus succinogenes*	McKinlay et al. (2007)
4	0.62	1.35	34	6.7	anaerobic	37	*Actinobacillus succinogenes*	Corona-González et al. (2008)
5	0.79	0.79	16	6.2	anaerobic	NA	*Anaerobiospirillum succiniproducens*	Samuelov et al. (1991)
6	0.87	1.93	44	6.1	anaerobic	39	*Anaerobiospirillum succiniproducens*	Glassner and Datta (1992)
7	0.88	0.87	33	6.4	anaerobic	39	*Anaerobiospirillum succiniproducens*	Datta (1992)
8	0.66	0.77	34	6.2	anaerobic	39	*Anaerobiospirillum succiniproducens*	Guettler and Jain (1996)
9	0.70	0.78	32	6.2	anaerobic	39	*Anaerobiospirillum succiniproducens*	Guettler and Jain (1996)
10	0.99	1.20	32	6.0	anaerobic	39	*Anaerobiospirillum succiniproducens*	Nghiem et al. (1997)
11	0.86	1.80	34	6.5	anaerobic	39	*Anaerobiospirillum succiniproducens*	Lee et al. (1999)
12	0.89	10.40	83	6.4	anaerobic	39	*Anaerobiospirillum succiniproducens*	Meynial-Salles et al. (2008)
13	0.71	3.30	14	6.5	anaerobic	39	*Mannheimia succiniproducens*	Lee et al. (2008)
14	0.70	1.87	14	6.5	anaerobic	39	*Mannheimia succiniproducens*	Lee et al. (2002)
15	0.76	1.80	52	6.5	anaerobic	39	*Mannheimia succiniproducens*	Lee et al. (2006)
16	0.59	1.75	11	6.5	anaerobic	39	*Mannheimia succiniproducens*	Song et al. (2007)
17	0.71	1.29	13	6.5	anaerobic	39	*Mannheimia succiniproducens*	Oh et al. (2008)
18	0.29	1.56	5	6.5	anaerobic	39	*Mannheimia succiniproducens*	Oh et al. (2008)
19	0.28	1.07	11	6.5	anaerobic	39	*Mannheimia succiniproducens*	Oh et al. (2008)
20	0.10	1.05	4	6.5	anaerobic	39	*Mannheimia succiniproducens*	Oh et al. (2008)
21	0.54	1.67	10	6.5	anaerobic	39	*Mannheimia succiniproducens*	Song et al. (2008)
22	0.29	0.59	11	6.0	anaerobic	NA	*Escherichia coli*	Millard et al. (1996)
23	0.42	0.43	4	6.4	anaerobic	37	*Escherichia coli*	Gokarn et al. (1998)
24	0.15	0.14	2	7.0	anaerobic	37	*Escherichia coli*	Gokarn et al. (2000)
25	0.45	0.16	5	7.0	aerobic	37	*Escherichia coli*	Lin et al. (2005d)
26	0.62	0.72	58	7.0	aerobic	37	*Escherichia coli*	Lin et al. (2005c)
27	0.71	0.14	8	7.0	aerobic	37	*Escherichia coli*	Lin et al. (2005b)
28	0.59	0.70	7	7.0	aerobic	37	*Escherichia coli*	Lin et al. (2005a)
29	1.06	0.42	40	7.0	anaerobic	37	*Escherichia coli*	Sánchez et al. (2005)
30	0.20	0.03	2	7.0	anaerobic	37	*Escherichia coli*	Lee et al. (2005)
31	1.10	0.42	40	7.0	anaerobic	37	*Escherichia coli*	Ka-Yiu et al. (2007)
32	0.90	0.90	87	7.0	anaerobic	37	*Escherichia coli*	Jantama et al. (2008)
33	1.10	1.30	98	7.0	anaerobic	37	*Escherichia coli*	Vemuri et al. (2002)
34	0.92	3.17	146	6.0	NA	NA	*Corneybacterium glutamicum*	Okino et al. (2008)
35	0.19	3.80	23	7.5	aerobic	NA	*Corneybacterium glutamicum*	Okino et al. (2005)

C Case study parameters

Table C.21: Summary of all literature-known data-sets for the production of succinic acid.

product	T_{melt} [°C]	T_B [°C]	c_p [$\frac{kJ}{kgK}$]	ρ [$\frac{g}{L}$]	h_{comb} [$\frac{kJ}{mol}$]	h_{fus} [$\frac{kJ}{mol}$]	h_{LV} [$\frac{kJ}{mol}$]	pKs_1	pKs_2	pKs_3	Sol_1 [$\frac{g}{L}$]	Sol_2 [1/°C]	$T_{Sol,low}$ [°C]	$T_{Sol,up}$ [°C]
ethanol	-114	78	2.4	790	1278	6	36	-	-	-	-	-	-	-
2,3-butanediol	36	182	2.6	994	2342	9	53	-	-	-	-	-	-	-
1,3-propanediol	-26	213	2.4[1]	1054	1745	14	57	-	-	-	-	-	-	-
itaconic acid	165	330	1.3[2]	1632	1967	24	60	3.85	5.45	-	31.691	0.0455	5	72
succinic acid	188	235	1.2	1560	1478	33	59	4.20	5.64	-	33.015	0.0353	23	174
fumaric acid	286	290	1.3	1635	1363	27	59	3.02	4.38	-	2.2936	0.0391	5	80
lactic acid	17	122	2.1[3]	1206	1294	14	56	3.86	-	-	-	-	-	-
HPA	25	280	2.1[3]	1283	1302	17	56	4.51	-	-	-	-	-	-
malic acid	132	306	1.3[2]	1601	1317	25	75	3.22	4.70	-	855.65	0.0189	5	65

Table C.22: References for experimental solubilities. The data-set used for regression is marked in bold.

itaconic acid	succinic acid	fumaric acid	malic acid
Krivankova et al. (1992)	Apelblat and Manzurola (1987)	Weiss and Downs (1923)	**Apelblat and Manzurola (1987)**
Apelblat and Manzurola (1997)	**Lin et al. (2007)**	**Lange and Sinks (1930)**	Apelblat et al. (1995)
Kuenz (2008)	Londono (2010)	Dang et al. (2009)	Daneshfar et al. (2012)
Urbanus et al. (2012)	Stephen and Stephen (2013)		Yuan et al. (2014)

[1] No component data available and, thus, cp value of 1,2-propanediol assumed.
[2] No component data available and, thus, cp value of succinic acid assumed.
[3] No component data available and, thus, cp value of propionic acid assumed.

C.9 Fermentation data

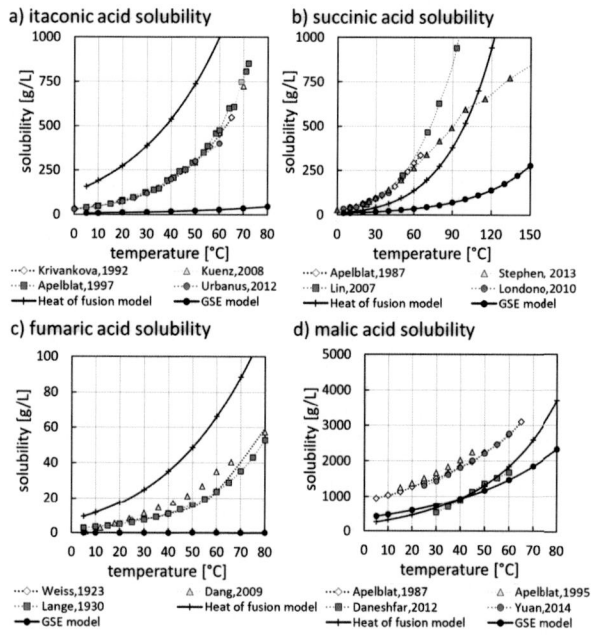

Figure C.3: An overview of experimental data and predictive models for itaconic (a), succinic (b), fumaric (c) and malic acid (d) solubility in water.

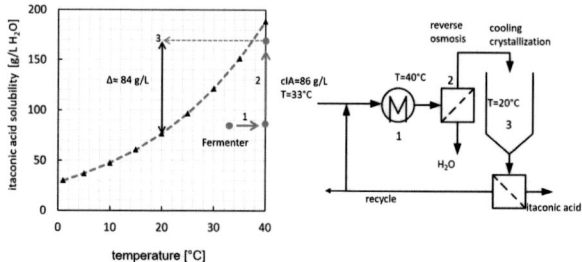

Figure C.4: Exemplary flowsheet for a final purification consisting of RO and cooling crystallization for the production of itaconic acid.

169

C.10 Conceptual design data

The minimum selling price MSP,

$$MSP = \frac{p}{\dot{m}}(\frac{ir}{1-(1+ir)^{-n}}IC + \sum_i cost_{raw\ materials,i} + \sum_i cost_{utility,i} + cost_{enz} + cost_{DS}),$$
(C.4)

is determined, based on the sum of the annualized investment cost, and the cost for raw materials, utility, enzymes enz and downstream processing DS. The required price parameters are given in Table . The IC are calculated using the approach proposed by Guthrie (1969). The price correlations are derived using an apparatus characteristic size, parameters for the operating conditions MPF, type of construction MF, and are updated to the recent year utilizing the Marshall-Swift-Index MS. For the heat exchanger and the plug flow reactor PFR, the following correlation is applied,

$$IC_{HEX} = \frac{MS_{2010}}{MS_{1969}} \cdot 5000\$ \cdot (\frac{Area_{HEX}[ft^2]}{400ft^2})^{0.77} \cdot (MPF_{HEX} + MF_{HEX} - 1),$$
(C.5)

which utilizes the shell surface area as main characteristic size. Herein, the PFR is considered as heat exchanger without shell assuming a minimal pipe velocity of 0,5 m/s and a maximal pipe diameter of 6 cm. Based on the geometry and the volume at the optimal operating point, the number of pipes is determined. The IC of a CSTR are derived using the cost correlation for a pressure vessel pv and in addition for a heat exchanger. Herein, the IC for a pressure vessel utilize the height H and diameter dia of a CSTR reactor according to

$$IC_{PV} = \frac{MS_{2010}}{MS_{1969}} \cdot 1000\$ \cdot (\frac{dia_{PV}[ft^2]}{3ft^2})^{1.05} \cdot (\frac{H_{PV}[ft^2]}{4ft^2})^{0.81} \cdot (MPF_{PV} + MF_{PV} - 1).$$
(C.6)

Table C.23: Price parameters

	Price	Reference
Biomass	90 €/ t	Voll (2014)
Sulphuric Acid (96 %)	75 $ / t	ICIS (2016)
Ammonia	180 $ / t	ICIS (2016)
Process water	0.14 €/ t	Poth (2013)
enzymes	3958 €/ t	Johnson (2016)
electricity	0.05 €/ kWh	Poth (2013)
steam 133 °C	0.02 €/ kWh	Skiborowski et al. (2015)
cooling water 15 °C	0.05 €/ t	Skiborowski et al. (2015)

C.10 Conceptual design data

Table C.24: Overview of references utilized for the conceptual design

process step	reference	description
pretreatment	Balat (2011)	beech wood composition as feed
	Lavarack et al. (2002)	kinetics for dilute acid pretreatment
enzymatic hydrolysis	Kadam et al. (2004)	kinetics utilize an enzyme mixture of endo- and exoglucanases
	Kadam et al. (2004)	Langmuir isotherm for enzyme-substrate adsorption
	Kadam et al. (2004)	Arrhenius approach for temperature dependency
fermentation	Krishnan et al. (1999)	reaction rates for co-fermentation
SSCF	Morales-Rodriguez et al. (2011)	inhibitory term for ethanol on cellulose hydrolysis to cellubiose
solid-liquid separator	Assumption	90% separation of solids
	Assumption	liquid content of solid stream 50%
reactor model	Recker (2017)	reactor model for all reaction steps
beer column	Bausa et al. (1998)	energy requirement
purification	Skiborowski (2014)	distillation coupled to pervaporation

Table C.25: Model comparison.

	RNFA	PNFA	Rigorous model
Level of detail of reactions	yields	yields	reaction kinetics
Pretreatment	Y=0.97	Y=0.97 dilute acid	
Enzymatic hydrolysis	yield, no enzyme cost	yield, no enzyme cost	kinetics, enzyme cost
Downstream	ideal	pressure-swing distillation	distillation + pervaporation
Investment costs	Empirical	Empirical	Physical sizing
Nonlinearity	low, only IC correlation	low, only IC correlation	high, difficult to solve

Bibliography

Abdehagh, N., Tezel, F. H., and Thibault, J. (2014). Separation techniques in butanol production: Challenges and developments. *Biomass and Bioenergy*, 60:222–246.

Ahlgren, S., Björklund, A., Ekman, A., Karlsson, H., Berlin, J., Börjesson, P., Ekvall, T., Finnveden, G., Janssen, M., and Strid, I. (2015). Review of methodological choices in LCA of biorefinery systems - key issues and recommendations. *Biofuels, Bioproducts and Biorefining*, 9(5):606–619.

Ahmad, S. (1985). *Heat Exchanger Networks: Cost Tradeoffs in Energy and Capital*. PhD thesis, University of Manchester, United Kingdom.

Al-Shaal, M. G., Dzierbinski, A., and Palkovits, R. (2014). Solvent-free gamma-valerolactone hydrogenation to 2-methyltetrahydrofuran catalysed by Ru/C: a reaction network analysis. *Green Chemistry*, 16(3):1358–1364.

Albrecht, T., Papadokonstantakis, S., Sugiyama, H., and Hungerbühler, K. (2010). Demonstrating multi-objective screening of chemical batch process alternatives during early design phases. *Chemical Engineering Research and Design*, 88(5-6):529–550.

Alibaba (2016). Professional supply intermediate ethyl levulinate CAS:539-88-8, https://www.alibaba.com/product-detail/Professional-supply-Intermediate-Ethyl-levulinate-CAS_1529946161.html, accessed November 23, 2016.

Alonso, D. M., Bond, J. Q., and Dumesic, J. A. (2010). Catalytic conversion of biomass to biofuels. *Green Chemistry*, 12(9):1493.

American Chemical Society (2015). SciFinder - The choice for chemistry research, http://www.cas.org/products/scifinder, accessed July 23, 2015.

Anastas, P. T. and Warner, J. C. (1998). *Green chemistry: Theory and practice*. Oxford University Press, Oxford and New York.

Anastas, P. T. and Zimmerman, J. B. (2003). Design through the 12 principles of green engineering. *Environmental Science & Technology*, 37(5):95A–101A.

173

Bibliography

Andiappan, V., Ko, A. S., Lau, V. W., Ng, L. Y., Ng, R. T., Chemmangattuvalappil, N. G., and Ng, D. K. (2015). Synthesis of sustainable integrated biorefinery via reaction pathway synthesis: Economic, incremental enviromental burden and energy assessment with multiobjective optimization. *AIChE Journal*, 61(1):132–146.

Andraos, J. (2005). Unification of reaction metrics for green chemistry II: Evaluation of named organic reactions and application to reaction discovery. *Organic Process Research & Development*, 9(4):404–431.

Angus-Hankin, C., Stokes, B., and Twaddle, A. (1995). The transportation of fuelwood from forest to facility. *Biomass and Bioenergy*, 9(1-5):191–203.

Anhui Jinao Chemical Co. Ltd. (2016). FOB price for Dimethyl Sulfoxide, http://www.alibaba.com/product-detail/Dimethyl-Sulfoxide-DMSO_1180950251.html?spm=a2700.7724857.29.3.owKOBk&s=p, accessed January 8, 2016.

Apelblat, A., Dov, M., Wisniak, J., and Zabicky, J. (1995). The vapour pressure of water over saturated aqueous solutions of malic, tartaric, and citric acids, at temperatures from 288 K to 323 K. *The Journal of Chemical Thermodynamics*, 27(1):35–41.

Apelblat, A. and Manzurola, E. (1987). Solubility of oxalic, malonic, succinic, adipic, maleic, malic, citric, and tartaric acids in water from 278.15 to 338.15 K. *The Journal of Chemical Thermodynamics*, 19(3):317–320.

Apelblat, A. and Manzurola, E. (1997). Solubilities of L-aspartic, DL-aspartic, DL-glutamic, p-hydroxybenzoic, o-anisic, p-anisic, and itaconic acids in water from T=278 K to T=345 K. *The Journal of Chemical Thermodynamics*, 29(12):1527–1533.

Arab, A. B. M. (2013). *Indicator-based Enhancement of "Green Chemistry and Engineering Principles" in Chemical Process Design*. PhD thesis, ETH Zurich, Switzerland.

Aral Service (2016). Preiszusammensetzung am Beispiel von Super E10, https://service.aral.de/preisbildung-und-struktur, accessed November, 10th, 2016.

Archambault-Léger, V., Losordo, Z., and Lynd, L. R. (2015). Energy, sugar dilution, and economic analysis of hot water flow-through pre-treatment for producing biofuel from sugarcane residues. *Biofuels, Bioproducts and Biorefining*, 9(1):95–108.

Atkins, P. W. and de Paula, J. (2014). *Atkins' Physical Chemistry*. University Press, Oxford, 10th edition.

Bibliography

Augé, J. (2008). A new rationale of reaction metrics for green chemistry. Mathematical expression of the environmental impact factor of chemical processes. *Green Chemistry*, 10(2):225–231.

Avraamidou, S. and Pistikopoulos, E. N. (2017). A Multiparametric Mixed-integer Bilevel Optimization Strategy for Supply Chain Planning Under Demand Uncertainty. *IFAC-PapersOnLine*, 50(1):10178–10183.

Badarinarayana, V., Rodwogin, M. D., Mullen, B. D., Purtle, I., and Molitor, E. J. (2014). Process to prepare levulinic acid, WO Patent 2014/189991.

Balat, M. (2011). Production of bioethanol from lignocellulosic materials via the biochemical pathway: A review. *Energy Conversion and Management*, 52(2):858–875.

Banimostafa, A., Papadokonstantakis, S., and Hungerbühler, K. (2012). Evaluation of EHS hazard and sustainability metrics during early process design stages using principal component analysis. *Process Safety and Environmental Protection*, 90(1):8–26.

Banimostafa, A., Papadokonstantakis, S., and Hungerbühler, K. (2015). Retrofit design of a pharmaceutical batch process considering "green chemistry and engineering" principles. *AIChE Journal*, 61(10):3423–3440.

Bao, B., Ng, D. K., Tay, D. H., Jiménez-Gutiérrez, A., and El-Halwagi, M. M. (2011). A shortcut method for the preliminary synthesis of process-technology pathways: An optimization approach and application for the conceptual design of integrated biorefineries. *Computers & Chemical Engineering*, 35(8):1374–1383.

Bartoli, G., Bosco, M., Carlone, A., Dalpozzo, R., Marcantoni, E., Melchiorre, P., and Sambri, L. (2007). Reaction of dicarbonates with carboxylic acids catalyzed by weak lewis acids: General method for the synthesis of anhydrides and esters. *Synthesis*, 22:3489–3496.

Bausa, J., Watzdorf, R. v., and Marquardt, W. (1998). Shortcut methods for nonideal multicomponent distillation: I. Simple columns. *AIChE Journal*, 44(10):2181–2198.

Beckmann, N. (2015). Concept development for an early stage process analysis of reaction pathways in a biorefinery. *Bachelor thesis (Supervisor: Ulonska, K.)*, Process Systems Engineering, RWTH Aachen University, Germany.

Benalcázar, E. A., Deynoot, B. G., Noorman, H., Osseweijer, P., and Posada, J. A. (2017). Production of bulk chemicals from lignocellulosic biomass via thermochemical conversion and syngas fermentation: A comparative techno-economic and environmental assessment of different site-specific supply chain configurations. *Biofuels, Bioproducts and Biorefining*, 11(5):861–886.

Bertók, B. and Fan, L. T. (2013). Review of methods for catalytic reaction-pathway identification at steady state. *Current Opinion in Chemical Engineering*, 2(4):487–494.

Bertran, M.-O., Frauzem, R., Sanchez-Arcilla, A.-S., Zhang, L., Woodley, J. M., and Gani, R. (2017). A generic methodology for processing route synthesis and design based on superstructure optimization. *Computers & Chemical Engineering*.

Besson, M., Gallezot, P., and Pinel, C. (2014). Conversion of biomass into chemicals over metal catalysts. *Chemical Reviews*, 114(3):1827–1870.

Biegler, L. T., Grossmann, I. E., and Westerberg, A. W. (1997). *Systematic Methods of Chemical Process Design: Prentice Hall international series in the physical and chemical engineering sciences*. Prentice Hall International Series in the Physical and Chemical Engineering Sciences. Prentice Hall PTR, Upper Saddle River, N.J.

Black, M., Sadhukhan, J., Day, K., Drage, G., and Murphy, R. (2016). Developing database criteria for the assessment of biomass supply chains for biorefinery development. *Chemical Engineering Research and Design*, 107:253–262.

Bollas, G. M., Barton, P. I., and Mitsos, A. (2009). Bilevel optimization formulation for parameter estimation in vapor–liquid(–liquid) phase equilibrium problems. *Chemical Engineering Science*, 64(8):1768–1783.

Bomheur, M. (2013). Prozess- und Kostenvergleich zur Herstellung von Butanol aus petrochemischen und nachwachsenden Rohstoffen. *Student thesis (Supervisors: Ulonska, K., Dahmen, M.)*, Process Systems Engineering, RWTH Aachen University, Germany.

Börjesson, P. (1996). Energy analysis of biomass production and transportation. *Biomass and Bioenergy*, 11(4):305–318.

Börjesson, P. and Gustavsson, L. (1996). Regional production and utilization of biomass in Sweden. *Energy*, 21(9):747–764.

Boudart, M. (1995). Turnover rates in heterogeneous catalysis. *Chemical Reviews*, 95(3):661–666.

Bridgwater, A. V. and Mumford, C. J. (1979). *Waste Recycling and Pollution Control Handbook*. Van Nostrand Reinhold environmental engineering series. Van Nostrand Reinhold, New York.

British Petroleum (2018). Answer to "Registration of isobutanol as a gasoline additive: opportunity for public comment", https://www.regulations.gov/document?D=EPA-HQ-OAR-2018-0131-0001, accessed June, 05, 2018.

Brodeur, G., Yau, E., Badal, K., Collier, J., Ramachandran, K. B., and Ramakrishnan, S. (2011). Chemical and physicochemical pretreatment of lignocellulosic biomass: A review. *Enzyme Research*, 2011(4999):1–17.

Broeren, M. L., Zijp, M. C., Waaijers-van der Loop, Susanne L., Heugens, E. H., Posthuma, L., Worrell, E., and Shen, L. (2017). Environmental assessment of biobased chemicals in early-stage development: A review of methods and indicators. *Biofuels, Bioproducts and Biorefining*, 11(4):701–718.

Brown, S. H., Bashkirova, L., Berka, R., Chandler, T., Doty, T., McCall, K., McCulloch, M., McFarland, S., Thompson, S., Yaver, D., and Berry, A. (2013). Metabolic engineering of *Aspergillus oryzae* NRRL 3488 for increased production of L-malic acid. *Applied Microbiology and Biotechnology*, 97(20):8903–8912.

Brown, T. R., Brown, R. C., and Estes, V. (2015). Commerical-scale production of lignocellulosic biofuels. *Chemical Engineering Progress*, (3):62–64.

Bundesamt für Kartographie und Geodäsie (2016). Verwaltungsgebiete 1:250 000, http://www.geodatenzentrum.de/geodaten/gdz_rahmen.gdz_div?gdz_spr=deu&gdz _akt_zeile=5&gdz_anz_zeile=1&gdz_unt_zeile=13&gdz_user_id=0, accessed May, 31, 2016.

Cambero, C. and Sowlati, T. (2014). Assessment and optimization of forest biomass supply chains from economic, social and environmental perspectives – A review of literature. *Renewable and Sustainable Energy Reviews*, 36:62–73.

Cambero, C., Sowlati, T., and Pavel, M. (2016). Economic and life cycle environmental optimization of forest-based biorefinery supply chains for bioenergy and biofuel production. *Chemical Engineering Research and Design*, 107:218–235.

Cao, N., Du, J., Gong, C., and Tsao, G. (1996). Simultaneous production and recovery of fumaric acid from immobilized *Rhizopus oryzae* with a rotary biofilm contactor and an adsorption column. *Applied and Environmental Microbiology*, 62(8):2926–2931.

Caspeta, L. and Nielsen, J. (2013). Economic and environmental impacts of microbial biodiesel. *Nature Biotechnology*, 31(9):789–793.

Celebi, A. D., Ensinas, A. V., Sharma, S., and Maréchal, F. (2017). Early-stage decision making approach for the selection of optimally integrated biorefinery processes. *Energy*, 137:908–916.

Chang, Y.-C., Lee, W.-J., Lin, S.-L., and Wang, L.-C. (2013). Green energy: Water-containing acetone–butanol–ethanol diesel blends fueled in diesel engines. *Applied Energy*, 109:182–191.

Cheali, P., Gernaey, K. V., and Sin, G. (2014). Toward a computer-aided synthesis and design of biorefinery networks: Data collection and management using a generic modeling approach. *ACS Sustainable Chemistry & Engineering*, 2(1):19–29.

Cheali, P., Gernaey, K. V., and Sin, G. (2015a). Uncertainties in early-stage capital cost estimation of process design - a case study on biorefinery design. *Frontiers in Energy Research*, 3:1–13.

Cheali, P., Posada, J. A., Gernaey, K. V., and Sin, G. (2015b). Upgrading of lignocellulosic biorefinery to value-added chemicals: Sustainability and economics of bioethanol-derivatives. *Biomass and Bioenergy*, 75:282–300.

Cherubini, F., Strømman, A. H., and Ulgiati, S. (2011). Influence of allocation methods on the environmental performance of biorefinery products—A case study. *Resources, Conservation and Recycling*, 55(11):1070–1077.

Ciric, A. and Floudas, C. (1991). Heat exchanger network synthesis without decomposition. *Computers & Chemical Engineering*, 15(6):385–396.

Clausen, U., Langkau, S., Goedicke, I., and Lier, S. (2015). Standort- und Netzwerkplanung für modulare Containeranlagen in der Prozessindustrie. *Chemie Ingenieur Technik*, 87(10):1365–1372.

Climent, M. J., Corma, A., and Iborra, S. (2014). Conversion of biomass platform molecules into fuel additives and liquid hydrocarbon fuels. *Green Chemistry*, 16(2):516–547.

Coker, A. K. and Ludwig, E. E. (2007). *Ludwig's Applied Process Design for Chemical and Petrochemical Plants*. Elsevier Gulf Professional Pub., Amsterdam and Boston, 4th edition.

Constable, D. J. C., Curzons, A. D., and Cunningham, V. L. (2002). Metrics to 'green' chemistry-which are the best? *Green Chemistry*, 4(6):521–527.

Corona-González, R. I., Bories, A., González-Álvarez, V., and Pelayo-Ortiz, C. (2008). Kinetic study of succinic acid production by *Actinobacillus succinogenes* ZT-130. *Process Biochemistry*, 43(10):1047–1053.

Čuček, L., Martín, M., Grossmann, I. E., and Kravanja, Z. (2014). Multi-period synthesis of optimally integrated biomass and bioenergy supply network. *Computers & Chemical Engineering*, 66:57–70.

Cui, J., Tan, J., Deng, T., Cui, X., Zheng, H., Zhu, Y., and Li, Y. (2015). Direct conversion of carbohydrates to gamma-valerolactone facilitated by a solvent effect. *Green Chemistry*, 17(5):3084–3089.

Curzons, A. D., Mortimer, D. N., Constable, D. J. C., and Cunningham, V. L. (2001). So you think your process is green, how do you know? — Using principles of sustainability to determine what is green – a corporate perspective. *Green Chemistry*, 3(1):1–6.

Cysewski, G. R. and Wilke, C. R. (1977). Rapid ethanol fermentations using vacuum and cell recycle. *Biotechnology and Bioengineering*, 19:1125–1143.

Dahmen, M. and Marquardt, W. (2016). Model-Based Design of Tailor-Made Biofuels. *Energy & Fuels*, 30(2):1109–1134.

Daneshfar, A., Baghlani, M., Sarabi, R. S., Sahraei, R., Abassi, S., Kaviyan, H., and Khezeli, T. (2012). Solubility of citric, malonic, and malic acids in different solvents from 303.2 to 333.2 K. *Fluid Phase Equilibria*, 313:11–15.

Dang, L., Du, W., Black, S., and Wei, H. (2009). Solubility of fumaric acid in propan-2-ol, ethanol, acetone, propan-1-ol, and water. *Journal of Chemical & Engineering Data*, 54(11):3112–3113.

Darkwah, K., Knutson, B. L., and Seay, J. R. (2018). A perspective on challenges and prospects for applying process systems engineering tools to fermentation-based biorefineries. *ACS Sustainable Chemistry & Engineering*, 6(3):2829–2844.

Datta, R. (1992). Process for the production of succinic acid by anaerobic fermentation, US Patent No. 5,143,833.

Dean, J. A. and Lange, N. A. (1999). *Lange's handbook of chemistry*. McGraw-Hill handbooks. McGraw-Hill, New York, NY [u.a.], 15. ed. edition.

del Rio-Chanona, E. A., Zhang, D., and Shah, N. (2018). Sustainable biopolymer synthesis via superstructure and multiobjective optimization. *AIChE Journal*, 64(1):91–103.

Demirbas, A. (2011). Competitive liquid biofuels from biomass. *Applied Energy*, 88(1):17–28.

Demirel, S. E., Li, J., and Hasan, F. M. (2017). Systematic process intensification using building blocks. *Computers & Chemical Engineering*, 105:2–38.

Deutsches Zentrum für Luft- und Raumfahrt (2016). Corine Land Cover 2006, http://www.corine.dfd.dlr.de, accessed May, 31, 2016.

Diederichsen, C. (1991). *Referenzumgebungen zur Berechnung der chemischen Exergie*. Fortschritt-Ber. VDI. Reihe 19, Nr. 50. Düsseldorf, VDI-Verlag.

Dieter, M., Englert, H., and Klein, M. (2001). Abschätzung des Rohholzpotentials für dieenergetische Nutzung in der Bundesrepublik Deutschland, Release date 2001, http://literatur.thuenen.de/digbib_extern/bitv/dk040192.pdf, accessed May 05, 2016.

Dortmund Data Bank Software & Separation Technology GmbH (2016). Online Group Assignment for UNIFAC and PSRK, accessed October 10, 2016.

Douglas, J. M. (1985). A Hierarchical Decision Procedure for Process Synthesis. *AIChE Journal*, 31(3):353–362.

Douglas, J. M. (2000). *Conceptual Design of Chemical Processes*. McGraw-Hill international editions. McGraw-Hill, New York, N.Y., 4th print edition.

Dufal, S., Papaioannou, V., Sadeqzadeh, M., Pogiatzis, T., Chremos, A., Adjiman, C. S., Jackson, G., and Galindo, A. (2014). Prediction of thermodynamic properties and phase behavior of fluids and mixtures with the SAFT-gamma Mie group-contribution equation of state. *Journal of Chemical & Engineering Data*, 59(10):3272–3288.

Dugar, D. and Stephanopoulos, G. (2011). Relative potential of biosynthetic pathways for biofuels and bio-based products. *Nature Biotechnology*, 29(12):1074–1078.

Dumesic, J. A., Serrano-Ruiz, J. C., and West, R. M. (2010). Catalytic conversion of cellulose to liquid hydrocarbon fuels by progressive removal of oxygen to facilitate separation processes and achieve high selectivities, US Patent No. 2010/0324310 A1.

Ebert, B. E. (2014). Theoretical fermentation yields. *Personal communication*, Institute of Applied Microbiology, Aachen, Germany.

Eckert, F. and Klamt, A. (2002). Fast solvent screening via quantum chemistry: COSMO-RS approach. *AIChE Journal*, 48(2):369–385.

Edwards, R., Larivé, J.-F., Rickeard, D., and Weindorf, W. (2014). Well-to-tank report version 4.a, JEC Well-to-wheels analysis, https://iet.jrc.ec.europa.eu/about-jec/sites/iet.jrc.ec.europa.eu.about-jec/files/documents/report_2014/wtt_report_v4a.pdf, Release date 2014, accessed April 03,2018: Technical Reports.

Eissen, M. and Metzger, J. O. (2002). Environmental performance metrics for daily use in synthetic chemistry. *Chemistry - A European Journal*, 8(16):3581–3585.

Ekşioğlu, S. D., Acharya, A., Leightley, L. E., and Arora, S. (2009). Analyzing the design and management of biomass-to-biorefinery supply chain. *Computers & Industrial Engineering*, 57(4):1342–1352.

El-Halwagi, M. M., editor (2012). *Sustainable Design Through Process Integration: Fundamentals and Applications to Industrial Pollution Prevention, Resource Conservation, and Profitability Enhancement*. Butterworth-Heinemann, Boston, MA, 1 edition.

El-Halwagi, M. M. and Manousiouthakis, V. (1989). Synthesis of mass exchange networks. *AIChE Journal*, 35(8):1233–1244.

El-Halwagi, M. M., Srinivas, B., and Dunn, R. F. (1995). Synthesis of optimal heat-induced separation networks. *Chemical Engineering Science*, 50(1):81–97.

Elia, J. A., Baliban, R. C., Floudas, C. A., Gurau, B., Weingarten, M. B., and Klotz, S. D. (2013). Hardwood Biomass to Gasoline, Diesel, and Jet Fuel: 2. Supply Chain Optimization Framework for a Network of Thermochemical Refineries. *Energy & Fuels*, 27(8):4325–4352.

Elnaghy, M. A. and Megalla, S. E. (1975). Itaconic-acid production by a local strain of *Aspergillus terreus*. *European Journal of Applied Microbiology*, 1(2):159–172.

Elsevier Information Systems GmbH (2015). Reaxys (R) - Version 2.19790.2, accessed August 28, 2015.

Esri (2016). ArcMap Version 10.4, http://desktop.arcgis.com/de/arcmap/, accessed September 20, 2016.

Ezeji, T., Qureshi, N., and Blaschek, H. P. (2007). Butanol production from agricultural residues: Impact of degradation products on *Clostridium beijerinckii* growth and butanol fermentation. *Biotechnology and Bioengineering*, 97(6):1460–1469.

Federal Reserve Bank St. Louis (2017). Producer price index by commodity chemicals and allied products: Industrial chemicals, https://fred.stlouisfed.org/series/WPU061, accessed January, 5th, 2017.

Fenkl, M., Pechout, M., Vojtisek, M., Dančová, P., and Veselý, M. (2016). N-butanol and isobutanol as alternatives to gasoline: Comparison of port fuel injector characteristics. *EPJ Web of Conferences*, 114(7):02021.

Floudas, C., Ciric, A., and Grossmann, I. (1986). Automatic synthesis of optimum heat exchanger network configurations. *AIChE Journal*, 32(2):276–290.

Frenzel, P., Hillerbrand, R., and Pfennig, A. (2014). Exergetical evaluation of biobased synthesis pathways. *Polymers*, 6(2):327–345.

Front Research (2016). Global formic acid market to grow slower in the years to come - Release date March 2016, http://www.frontresearch.com/news/global-formic-acid-market-to-grow-slower-in-the-years-to-come/, accessed November 23, 2016.

Fu, M.-C., Shang, R., Huang, Z., and Fu, Y. (2014). Conversion of levulinate ester and formic acid into gammavalerolactone using a homogeneous iron catalyst. *Synlett*, 25(19):2748–2752.

GAMS Development Corporation (2012). GAMS – General Algebraic Modeling System, http://www.gams.com, accessed August, 10, 2012.

Gani, R., Hytoft, G., Jaksland, C., and Jensen, A. K. (1997). An integrated computer aided system for integrated design of chemical processes. *Computers & Chemical Engineering*, 21(10):1135–1146.

Gani, R., Jiménez-González, C., and Constable, D. J. (2005). Method for selection of solvents for promotion of organic reactions. *Computers & Chemical Engineering*, 29(7):1661–1676.

Gantner, P. (2015). Applicability of RNFA methodology for evaluation of process options based on large reaction network datasets. *Master thesis (Supervisors: Ulonska, K., Lapkin, A.)*, Process Systems Engineering, RWTH Aachen University, Germany.

Gao, J. and You, F. (2017). Modeling framework and computational algorithm for hedging against uncertainty in sustainable supply chain design using functional-unit-based life cycle optimization. *Computers & Chemical Engineering*, 107:221–236.

Gao, T., Wong, Y., Ng, C., and Ho, K. (2012). L-lactic acid production by *Bacillus subtilis* MUR1. *Bioresource Technology*, 121:105–110.

Garcia, D. J. and You, F. (2015a). Multiobjective optimization of product and process networks: General modeling framework, efficient global optimization algorithm, and case studies on bioconversion. *AIChE Journal*, 61(2):530–554.

Garcia, D. J. and You, F. (2015b). Supply chain design and optimization: Challenges and opportunities. *Computers & Chemical Engineering*, 81:153–170.

Gargalo, C. L., Carvalho, A., Gernaey, K. V., and Sin, G. (2016a). A framework for techno-economic & environmental sustainability analysis by risk assessment for conceptual process evaluation. *Biochemical Engineering Journal*, 116:146–156.

Gargalo, C. L., Cheali, P., Posada, J. A., Carvalho, A., Gernaey, K. V., and Sin, G. (2016b). Assessing the environmental sustainability of early stage design for bioprocesses under uncertainties: An analysis of glycerol bioconversion. *Journal of Cleaner Production*, 139:1245–1260.

Geilen, F. M. A., Engendahl, B., Harwardt, A., Marquardt, W., Klankermayer, J., and Leitner, W. (2010). Selective and flexible transformation of biomass-derived platform chemicals by a multifunctional catalytic system. *Angewandte Chemie International Edition*, 49(32):5510–5514.

Geilen, F. M. A., vom Stein, T., Engendahl, B., Winterle, S., Liauw, M. A., Klankermayer, J., and Leitner, W. (2011). Highly selective decarbonylation of 5-(hydroxymethyl)furfural in the presence of compressed carbon dioxide. *Angewandte Chemie International Edition*, 50(30):6831–6834.

Geraili, A. and Romagnoli, J. A. (2015). A multiobjective optimization framework for design of integrated biorefineries under uncertainty. *AIChE Journal*, 61(10):3208–3222.

Gerdes, A.-L. (2014). Ökonomische und ökologische Bewertung von Fermentationsprozessen. *Bachelor thesis (Supervisor: Ulonska, K.)*, Process Systems Engineering, RWTH Aachen University,Germany.

Gerrard, A. M. (2000). *Guide to Capital Cost Estimating*. Institution of Chemical Engineers, Rugby, Warwickshire, U.K, 4th ed edition.

Gevo (2016). Gevo working on bringing down isobutanol production costs to \$3 per gallon,http://www.biofuelsdigest.com/bdigest/2016/11/17/gevo-working-on-bringing-down-isobutanol-production-costs-to-3-per-gallon/, Release data 2016, accessed March 29, 2018: Company statement.

Ghaderi, H., Pishvaee, M. S., and Moini, A. (2016). Biomass supply chain network design: An optimization-oriented review and analysis. *Industrial Crops and Products*, 94:972–1000.

Giuliano, A., Poletto, M., and Barletta, D. (2016). Process optimization of a multi-product biorefinery: The effect of biomass seasonality. *Chemical Engineering Research and Design*, 107:236–252.

Glass, M., Aigner, M., Viell, J., Jupke, A., and Mitsos, A. (2017). Liquid-liquid equilibrium of 2-methyltetrahydrofuran/water over wide temperature range: Measurements and rigorous regression. *Fluid Phase Equilibria*, 433:212–225.

Glassner, D. A. and Datta, R. (1992). Process for the production and purification of succinic acid, US Patent No. 5,143,834.

Gokarn, R. R., Eiteman, M. A., and Altman, E. (1998). Expression of pyruvate carboxylase enhances succinate production in *Escherichia coli* without affecting glucose uptake. *Biotechnology Letters*, 20(8):795–798.

Bibliography

Gokarn, R. R., Eiteman, M. A., and Altman, E. (2000). Metabolic analysis of *Escherichia coli* in the presence and absence of the carboxylating enzymes *Phosphoenolpyruvate Carboxylase* and *Pyruvate Carboxylase*. *Applied and Environmental Microbiology*, 66(5):1844–1850.

Gong, Y., Lin, L., Shi, J., and Liu, S. (2010). Oxidative decarboxylation of levulinic acid by cupric oxides. *Molecules*, 15(11):7946–7960.

Graham-Rowe, D. (2011). Agriculture: Beyond food versus fuel. *Nature*, 474(7352):S6–S8.

Grand View Research (2016a). Global ethyllevulinate market - Release date March 2016, https://www.grandviewresearch.com/press-release/global-ethyl-levulinate-market, accessed November 23, 2016.

Grand View Research (2016b). Global furfural market - Release date March 2008, https://www.grandviewresearch.com/press-release/global-furfural-market, accessed November 23, 2016.

Grand View Research (2016c). Global isobutanol market - Release date March 2016, https://www.grandviewresearch.com/press-release/global-isobutanol-market, accessed November 23, 2016.

Grand View Research (2016d). Global levulinic acid market - Release date March 2015, https://www.grandviewresearch.com/press-release/global-levulinic-acid-market, accessed November 23, 2016.

Grand View Research (2016e). Global methyl ethyl ketone market - Release date August 2016, https://www.grandviewresearch.com/press-release/global-methyl-ethyl-ketone-mek-market, accessed November 23, 2016.

Green, E. M. (2011). Fermentative production of butanol–the industrial perspective. *Current Opinion in Biotechnology*, 22(3):337–343.

Grossmann, I. E. (1990). Mixed-integer nonlinear programming techniques for the synthesis of engineering systems. *Res. Eng. Des.*, 1:205–228.

Guettler, M. V. and Jain, M. K. (1996). Method for making succinic acid, *Anaerobiospirillum succiniproducens* variants for use in process and methods for obtaining variants, US Patent No. 5,521,075.

Guthrie, K. M. (1969). Capital Cost Estimating. *Chemical Engineering*, 76(6):114.

Halemane, K. P. and Grossmann, I. E. (1983). Optimal process design under uncertainty. *AIChE Journal*, 29(3):425–433.

Hamelinck, C. N., Suurs, R. A., and Faaij, A. P. (2005). International bioenergy transport costs and energy balance. *Biomass and Bioenergy*, 29(2):114–134.

Harwardt, A. and Marquardt, W. (2012). Heat-integrated distillation columns: Vapor recompression or internal heat integration? *AIChE Journal*, 58(12):3740–3750.

Hashim, H., Narayanasamy, M., Yunus, N. A., Shiun, L. J., Muis, Z. A., and Ho, W. S. (2017). A cleaner and greener fuel: Biofuel blend formulation and emission assessment. *Journal of Cleaner Production*, 146:208–217.

Hayes, D. J., Fitzpatrick, S., Hayes, M. H., and Ross, J. R. (2006). The Biofine Process: Production of levulinic acid, furfural and formic acid from lignocellulosic feedstocks. In Kamm, B., Gruber, P. R., and Kamm, M., editors, *Biorefineries*, pages 139–164. Wiley-VCH, Weinheim.

Hechinger, M., Voll, A., and Marquardt, W. (2010). Towards an integrated design of biofuels and their production pathways. *Computers & Chemical Engineering*, 34:1909–1918.

Heinzle, E., Weirich, D., Brogli, F., Hoffmann, V. H., Koller, G., Verduyn, M. A., and Hungerbühler, K. (1998). Ecological and economic objective functions for screening in integrated development of fine chemical processes. 1. Flexible and expandable framework using indices. *Industrial & Engineering Chemistry Research*, 37(8):3395–3407.

Henry, C. S., Broadbelt, L. J., and Hatzimanikatis, V. (2007). Thermodynamics-based metabolic flux analysis. *Biophysical Journal*, 92(5):1792–1805.

Henry, C. S., Broadbelt, L. J., and Hatzimanikatis, V. (2010). Discovery and analysis of novel metabolic pathways for the biosynthesis of industrial chemicals: 3-hydroxypropanoate. *Biotechnology and Bioengineering*, 106(3):462–473.

Hermann, B. G., Blok, K., and Patel, M. K. (2007). Producing bio-based bulk chemicals using industrial biotechnology saves energy and combats climate change. *Environmental Science & Technology*, 41(22):7915–7921.

Hermann, B. G. and Patel, M. (2007). Today's and tomorrow's bio-based bulk chemicals from white biotechnology: A techno-economic analysis. *Applied Biochemistry and Biotechnology*, 136:361–388.

Heuser, B., Laible, T., Jakob, M., Kremer, F., and Pischinger, S. (2014). C8-oxygenates for clean diesel combustion. *SAE Technical Paper*, 01(1253).

Hevekerl, A., Kuenz, A., and Vorlop, K.-D. (2014). Filamentous fungi in microtiter plates—an easy way to optimize itaconic acid production with *Aspergillus terreus*. *Applied Microbiology and Biotechnology*, 98(16):6983–6989.

Bibliography

Holtbruegge, J., Kuhlmann, H., and Lutze, P. (2014). Conceptual design of flow-sheet options based on thermodynamic insights for (reaction−)separation processes applying process intensification. *Industrial & Engineering Chemistry Research*, 53(34):140814103548004.

Hoppe, F., Heuser, B., Thewes, M., Kremer, F., Pischinger, S., Dahmen, M., Hechinger, M., and Marquardt, W. (2015). Tailor-made fuels for future engine concepts. *International Journal of Engine Research*, 17(1):16–27.

Huang, Y., Chen, C.-W., and Fan, Y. (2010). Multistage optimization of the supply chains of biofuels. *Transportation Research Part E: Logistics and Transportation Review*, 46(6):820–830.

Huang, Y., Li, Z., Shimizu, K., and Ye, Q. (2013). Co-production of 3-hydroxypropionic acid and 1,3-propanediol by *Klebseilla pneumoniae* expressing aldH under microaerobic conditions. *Bioresource Technology*, 128:505–512.

Huber, G. W., Iborra, S., and Corma, A. (2006). Synthesis of transportation fuels from biomass: chemistry, catalysts, and engineering. *Chemical Reviews*, 106(9):4044–4098.

Hudlicky, T., Frey, D. A., Koroniak, L., Claebo, C. D., and Brammer Jr., L. E. (1999). Toward a 'reagent-free' synthesis -tandem enzymatic and electrochemical methods for increased effective mass yield (EMY). *Green Chemistry*, 1:57–59.

Humbird, D., Davis, R., Tao, L., Kinchin, C., Hsu, D., and Aden, A. (2012). Process Design and Economics for Biochemical Conversion of Lignocellulosic Biomass to Ethanol: Dilute-Acid Pretreatment and Enzymatic Hydrolysis of Corn Stover, Release date 2011: Technical Report NREL/TP-5100-47764 National Renewable Energy Laboratory.

ICIS (2016). Oxo alcohols market coping with high propylene prices- Release date October 2014, http://www.icis.com/resources/news/2014/10/31/9834501/us-oxo-alcohols-market-coping-with-high-propylene-prices/, accessed November 23, 2016.

Ileleji, K. E., Sokhansanj, S., and Cundiff, J. S. (2010). Farm-gate to plant-gate delivery of lignocellulosic feedstocks from plant biomass for biofuel production. In Blaschek, H. P., Ezeji, T. C., and Scheffran, J., editors, *Biofuels from Agricultural Wastes and Byproducts*, pages 117–159. Wiley-Blackwell, Ames, Iowa.

Iman, R. L. and Helton, J. C. (1988). An Investigation of uncertainty and sensitivity analysis techniques for computer models. *Risk Analysis*, 8(1):71–90.

Institute for Occupational Safety and Health of the German Social Accident Insurance (2016). GESTIS Substance Database, http://gestis-en.itrust.de, accessed August 10, 2016.

International Institute for Sustainability Analysis and Strategy (2016). Global Emissions Model for integrated Systems (GEMIS), Version 4.94, http://www.iinas.org/gemis-de.html, accessed August 10, 2016.

ISO 14040 (2006). Environmental management – Life cycle assessment – Principles and framework.

Jain, N., Yang, G., Machatha, S. G., and Yalkowsky, S. H. (2006). Estimation of the aqueous solubility of weak electrolytes. *International journal of pharmaceutics*, 319(1-2):169–171.

Jakob, A., Grensemann, H., Lohmann, J., and Gmehling, J. (2006). Further development of modified UNIFAC (Dortmund): Revision and extension 5. *Industrial & Engineering Chemistry Research*, 45(23):7924–7933.

Jaksland, C. A., Gani, R., and Lien, K. M. (1995). Separation process design and synthesis based on thermodynamic insights. *Chemical Engineering Science*, 50(3):511–530.

Jankowski, M. D., Henry, C. S., Broadbelt, L. J., and Hatzimanikatis, V. (2008). Group contribution method for thermodynamic analysis of complex metabolic networks. *Biophysical Journal*, 95(3):1487–1499.

Jantama, K., Haupt, M. J., Svoronos, S. A., Zhang, X., Moore, J. C., Shanmugam, K. T., and Ingram, L. O. (2008). Combining metabolic engineering and metabolic evolution to develop nonrecombinant strains of *Escherichia coli* C that produce succinate and malate. *Biotechnology and bioengineering*, 99(5):1140–1153.

Ji, X.-J., Huang, H., and Ouyang, P.-K. (2011). Microbial 2,3-butanediol production: A state-of-the-art review. *Biotechnology Advances*, 29(3):351–364.

Jiangsu Senxuan Pharmaceutical And Chemical Co. Ltd. (2016). FOB Price for 1,4-Dioxane, http://www.alibaba.com/product-detail/1-4-Dioxane_966670201.html?spm=a2700.7724857.29.12.MU6DzM&s=p, accessed January 8, 2016.

Jimenez-Gonzalez, C., Ponder, C. S., Broxterman, Q. B., and Manley, J. B. (2011). Using the right green yardstick: Why process mass intensity is used in the pharmaceutical industry to drive more sustainable processes. *Organic Process Research & Development*, 15(4):912–917.

Joback, K. and Reid, R. (1987). Estimation of pure-component properties from group-contributions. *Chemical Engineering Communications*, 57(1-6):233–243.

Johnson, E. (2016). Integrated enzyme production lowers the cost of cellulosic ethanol. *Biofuels, Bioproducts and Biorefining*, 10(2):164–174.

Jonker, J., Junginger, H. M., Verstegen, J. A., Lin, T., Rodríguez, L. F., Ting, K. C., Faaij, A., and van der Hilst, F. (2016). Supply chain optimization of sugarcane first generation and eucalyptus second generation ethanol production in Brazil. *Applied Energy*, 173:494–510.

Ju, N. and Wang, S. S. (1986). Continuous production of itaconic acid by *Aspergillus terreus* immobilized in a porous disk bioreactor. *Applied Microbiology and Biotechnology*, 23(5):311–314.

Julis, J. and Leitner, W. (2012). Synthesis of 1-octanol and 1,1-dioctyl ether from biomass-derived platform chemicals. *Angewandte Chemie International Edition*, 51(34):8615–8619.

Ka-Yiu, S., Bennett, G. N., and Sanchez, A. (2007). Mutant *E.coli* strain with increased succinic acid production, US Patent No. 11/214,309.

Kadam, K. L., Rydholm, E. C., and McMillan, J. D. (2004). Development and validation of a kinetic model for enzymatic saccharification of lignocellulosic biomass. *Biotechnology progress*, 20(3):698–705.

Kamm, B., Gruber, P. R., and Kamm, M., editors (2006). *Biorefineries: Industrial processes and directions : status quo and future directions*. Wiley-VCH, Weinheim.

Kappler, G. O. (2008). *Systemanalytische Untersuchung zum Aufkommen und zur Bereitstellung von energetisch nutzbarem Reststroh und Waldrestholz in Baden-Württemberg*. PhD thesis, Forst- und Umweltwissenschaften, Albert-Ludwigs-Universität Freiburg, Germany.

Kaufman, A. S., Meier, P. J., Sinistore, J. C., and Reinemann, D. J. (2010). Applying life-cycle assessment to low carbon fuel standards—How allocation choices influence carbon intensity for renewable transportation fuels. *Energy Policy*, 38(9):5229–5241.

Kautola, H. (1990). Itaconic acid production from xylose in repeated-batch and continuous bioreactors. *Applied Microbiology and Biotechnology*, 33(1):7–11.

Kautola, H., Rymowicz, W., Linko, Y.-Y., and Linko, P. (1991). Itaconic acid production by immobilized *Aspergillus terreus* with varied metal additions. *Applied Microbiology and Biotechnology*, 35(2):154–158.

Kautola, H., Vahvaselkä, M., Linko, Y.-Y., and Linko, P. (1985). Itaconic acid production by immobilized *Aspergillus terreus* from xylose and glucose. *Biotechnology Letters*, 7(3):167–172.

Kautola, H., Vassilev, N., and Linko, Y.-Y. (1990). Continuous itaconic acid production by immobilized biocatalysts. *Journal of Biotechnology*, 13(4):315–323.

Kelloway, A. and Daoutidis, P. (2014). Process synthesis of biorefineries: optimization of biomass conversion to fuels and chemicals. *Industrial & Engineering Chemistry Research*, 53(13):5261–5273.

Kerschgens, B., Cai, L., Pitsch, H., Heuser, B., and Pischinger, S. (2016). Di- n -buthylether, n -octanol, and n -octane as fuel candidates for diesel engine combustion. *Combustion and Flame*, 163:66–78.

Khusnutdinov, R. I., Bayguzina, A. R., Gimaletdinova, L. I., and Dzhemilev, U. M. (2012). Intermolecular dehydration of alcohols by the action of copper compounds activated with carbon tetrabromide. Synthesis of ethers. *Russian Journal of Organic Chemistry*, 48(9):1191–1196.

Kim, J., Realff, M. J., and Lee, J. H. (2011a). Optimal design and global sensitivity analysis of biomass supply chain networks for biofuels under uncertainty. *Computers & Chemical Engineering*, 35(9):1738–1751.

Kim, J., Realff, M. J., Lee, J. H., Whittaker, C., and Furtner, L. (2011b). Design of biomass processing network for biofuel production using an MILP model. *Biomass and Bioenergy*, 35(2):853–871.

Kim, J., Sen, S. M., and Maravelias, C. T. (2013). An optimization-based assessment framework for biomass-to-fuel conversion strategies. *Energy & Environmental Science*, 6(4):1093.

Kim, S. and Dale, B. E. (2003). Cumulative energy and global warming impact from the production of biomass for biobased products. *Journal of Industrial Ecology*, 7(3-4):147–162.

Klamt, A. (2005). *COSMO-RS -From Quantum Chemistry to Fluid Phase Thermodynamics and Drug Design*. Elsevier, Amsterdam and Boston, 1st edition.

Klatt, M. (2013). Life Cycle Analyse und Kostenevaluierung zur Herstellung von Tetrahydrofuran. *Student thesis (Supervisor: Ulonska, K.)*, Process Systems Engineering, RWTH Aachen University,Germany.

Klein-Marcuschamer, D. and Blanch, H. W. (2013). Survival of the fittest: An economic perspective on the production of novel biofuels. *AIChE Journal*, 59(12):4454–4460.

Klein-Marcuschamer, D. and Blanch, H. W. (2015). Renewable fuels from biomass: Technical hurdles and economic assessment of biological routes. *AIChE Journal*, 61(9):2689–2701.

Klein-Marcuschamer, D., Oleskowicz-Popiel, P., Simmons, B. A., and Blanch, H. W. (2010). Technoeconomic analysis of biofuels: A wiki-based platform for lignocellulosic biorefineries. *Biomass and Bioenergy*, 34(12):1914–1921.

Klein-Marcuschamer, D., Oleskowicz-Popiel, P., Simmons, B. A., and Blanch, H. W. (2012). The challenge of enzyme cost in the production of lignocellulosic biofuels. *Biotechnology and Bioengineering*, 109(4):1083–1087.

Klement, T., Milker, S., Jäger, G., Grande, P. M., Domínguez de María, P., and Büchs, J. (2012). Biomass pretreatment affects *Ustilago maydis* in producing itaconic acid. *Microbial cell factories*, 11:43.

Kochs, A. (2015). Entwicklung von Verfahrensbewertungen zur Herstellung von Fermentationsprodukten in Bioraffinerien. *Bachelor thesis (Supervisor: Ulonska, K.,Schupp, T.)*, Process Systems Engineering, RWTH Aachen University and Fachhochschule Münster, Germany.

Kokossis, A., Tsakalova, M., and Pyrgakis, K. (2015). Design of integrated biorefineries. *Computers & Chemical Engineering*, 81:40–56.

Koller, G., Fischer, U., and Hungerbühler, K. (2000). Assessing safety, health, and environmental impact early during process development. *Industrial & Engineering Chemistry Research*, 39(4):960–972.

Kondili, E., Pantelides, C., and Sargent, R. (1993). A general algorithm for short-term scheduling of batch operations—I. MILP formulation. *Computers & Chemical Engineering*, 17(2):211–227.

Kong, L., Sen, S. M., Henao, C. A., Dumesic, J. A., and Maravelias, C. T. (2016). A superstructure-based framework for simultaneous process synthesis, heat integration, and utility plant design. *Computers & Chemical Engineering*, 91:68–84.

Kong, Q. and Shah, N. (2016). An optimisation-based framework for the conceptual design of reaction-separation processes. *Chemical Engineering Research and Design*, 113:206–222.

König, A. (2016). Design of optimal multi-product biorefineries considering the biomass supply chain. *Master thesis (Supervisor: Ulonska, K.)*, Process Systems Engineering, RWTH Aachen University,Germany.

König, A., Ulonska, K., Viell, J., and Mitsos, A. (2018). Systematic performance comparison of production pathways for renewable fuels from biomass, electricity, and combinations thereof. *In preparation*.

Kossack, S., Kraemer, K., Gani, R., and Marquardt, W. (2008). A systematic synthesis framework for extractive distillation processes. *Chemical Engineering Research and Design*, 86(7):781–792.

Kraemer, K., Kossack, S., and Marquardt, W. (2009). Efficient optimization-based design of distillation processes for homogeneous azeotropic mixtures. *Industrial & Engineering Chemistry Research*, 48(14):6749–6764.

Kreyenschulte, D., Emde, F., Regestein, L., and Büchs, J. (2016). Computational minimization of the specific energy demand of large-scale aerobic fermentation processes based on small-scale data. *Chemical Engineering Science*, 153:270–283.

Krishnan, M. S., Ho, N. W. Y., and Tsao, G. T. (1999). Fermentation kinetics of ethanol production from glucose and xylose by recombinant *Saccharomyces* 1400(pLNH33). *Applied Biochemistry and Biotechnology*, 78(1-3):373–388.

Krivankova, I., Marcisinova, M., and Sohnel, O. (1992). Solubility of itaconic and kojic acids. *Journal of Chemical & Engineering Data*, 37(1):23–24.

Kuenz, A. (2008). *Itaconsäureherstellung aus nachwachsenden Rohstoffen als Ersatz für petrochemisch hergestellte Acrylsäure*. PhD thesis, Fakultät für Lebenswissenschaften, Technische Universität Carolo-Wilhelmina zu Braunschweig, Braunschweig.

Kuenz, A., Gallenmüller, Y., Willke, T., and Vorlop, K.-D. (2012). Microbial production of itaconic acid: developing a stable platform for high product concentrations. *Applied Microbiology and Biotechnology*, 96(5):1209–1216.

Kuhlmann, H. and Skiborowski, M. (2017). Optimization-based approach to process synthesis for process intensification: General approach and application to ethanol dehydration. *Industrial & Engineering Chemistry Research*, 56(45):13461–13481.

Kumar, A., Sokhansanj, S., and Flynn, P. C. (2006). Development of a multicriteria assessment model for ranking biomass feedstock collection and transportation systems. In McMillan, J. D., Adney, W. S., Mielenz, J. R., and Klasson, K. T., editors, *Twenty-Seventh Symposium on Biotechnology for Fuels and Chemicals*, pages 71–87. Humana Press, Totowa, NJ.

Kumar, M., Goyal, Y., Sarkar, A., and Gayen, K. (2012). Comparative economic assessment of ABE fermentation based on cellulosic and non-cellulosic feedstocks. *Applied Energy*, 93:193–204.

Kumar, P., Barrett, D. M., Delwiche, M. J., and Stroeve, P. (2009). Methods for pretreatment of lignocellulosic biomass for efficient hydrolysis and biofuel production. *Industrial & Engineering Chemistry Research*, 48(8):3713–3729.

Kurzrock, T. and Weuster-Botz, D. (2010). Recovery of succinic acid from fermentation broth. *Biotechnology Letters*, 32(3):331–339.

Lange, J.-P. (2001). Fuels and chemicals manufacturing; Guidelines for understanding and minimizing the production costs. *Cattech*, 5(2):82–95.

Lange, J.-P. (2007). Lignocellulose conversion: an introduction to chemistry, process and economics. *Biofuels, Bioproducts and Biorefining*, 1(1):39–48.

Lange, N. A. and Sinks, M. H. (1930). The solubility, specific gravity and index of refraction of aqueous solutions of fumaric, maleid and L-malic acids. *Journal of the American Chemical Society*, 52(7):2602–2604.

Lanzafame, P., Temi, D., Perathoner, S., Centi, G., Macario, A., Aloise, A., and Giordano, G. (2011). Etherification of 5-hydroxymethyl-2-furfural (HMF) with ethanol to biodiesel components using mesoporous solid acidic catalysts. *Catalysis Today*, 175(1):435–441.

Lara, C. L. and Grossmann, I. E. (2016). Global Optimization for a Continuous Location-Allocation Model for Centralized and Distributed Manufacturing. *Computer Aided Chemical Engineering*, 38:1009–1014.

Lavarack, B. P., Griffin, G. J., and Rodman, D. (2002). The acid hydrolysis of sugarcane bagasse hemicellulose to produce xylose, arabinose, glucose and other products. *Biomass and Bioenergy*, 23(5):367–380.

Lee, P. C., Lee, S. Y., Hong, S. H., and Chang, H. N. (2002). Isolation and characterization of a new succinic acid-producing bacterium, *Mannheimia succiniciproducens* MBEL55E, from bovine rumen. *Applied Microbiology and Biotechnology*, 58(5):663–668.

Lee, P.-C., Sang-Yup, L., and Ho-Nam, C. (2008). Cell recycled culture of succinic acid-producing *Anaerobiospirillum succiniciproducens* using an internal membrane filtration system. *Journal of Microbiology and Biotechnology*, 18(7):1252–1256.

Lee, P. C., Woo Gi Lee,, G. L., Sunhoon, K., Sang Yup, L., and Ho Nam, C. (1999). Succinic acid production by *Anaerobiospirillum succiniciproducens*: effects of the H2/CO2 supply and glucose concentration. *Enzyme and Microbial Technology*, 24:549–554.

Lee, S. J., Lee, D.-Y., Kim, T. Y., Kim, B. H., Lee, J., and Lee, S. Y. (2005). Metabolic engineering of *Escherichia coli* for enhanced production of succinic acid, based on genome comparison and in silico gene knockout simulation. *Applied and Environmental Microbiology*, 71(12):7880–7887.

Lee, S. J., Song, H., and Lee, S. Y. (2006). Genome-based metabolic engineering of *Mannheimia succiniciproducens* for succinic acid production. *Applied and Environmental Microbiology*, 72(3):1939–1948.

Li, Y., Nithyanandan, K., Lee, T. H., Donahue, R. M., Lin, Y., Lee, C.-F., and Liao, S. (2016). Effect of water-containing acetone–butanol–ethanol gasoline blends on combustion, performance, and emissions characteristics of a spark-ignition engine. *Energy Conversion and Management*, 117:21–30.

Lin, H., Bennett, G. N., and San, K.-Y. (2005a). Chemostat culture characterization of *Escherichia coli* mutant strains metabolically engineered for aerobic succinate production: a study of the modified metabolic network based on metabolite profile, enzyme activity, and gene expression profile. *Metabolic Engineering*, 7(5-6):337–352.

Lin, H., Bennett, G. N., and San, K.-Y. (2005b). Fed-batch culture of a metabolically engineered *Escherichia coli* strain designed for high-level succinate production and yield under aerobic conditions. *Biotechnology and bioengineering*, 90(6):775–779.

Lin, H., Bennett, G. N., and San, K.-Y. (2005c). Genetic reconstruction of the aerobic central metabolism in *Escherichia coli* for the absolute aerobic production of succinate. *Biotechnology and bioengineering*, 89(2):148–156.

Lin, H., Bennett, G. N., and San, K.-Y. (2005d). Metabolic engineering of aerobic succinate production systems in *Escherichia coli* to improve process productivity and achieve the maximum theoretical succinate yield. *Metabolic Engineering*, 7(2):116–127.

Lin, H.-M., Tien, H.-Y., Hone, Y.-T., and Lee, M.-J. (2007). Solubility of selected dibasic carboxylic acids in water, in ionic liquid of [Bmim][BF4], and in aqueous [Bmim][BF4] solutions. *Fluid Phase Equilibria*, 253(2):130–136.

Ling, L. B. and Ng, T. K. (1989). Fermentation process for carboxylic acids, US Patent No. 4,877,731.

Linnhoff, B. (1993). Pinch analysis: a state-of-the-art overview. *Chemical Engineering Research and Design*, 71(5):503–522.

Linnhoff, B. and Ahmad, S. (1989). Supertargeting: Optimum Synthesis of Energy Management Systems. *Journal of Energy Resources Technology*, 111(3):121–130.

Liu, R., Chen, J., Huang, X., Chen, L., Ma, L., and Li, X. (2013). Conversion of fructose into 5-hydroxymethylfurfural and alkyl levulinates catalyzed by sulfonic acid-functionalized carbon materials. *Green Chemistry*, 15(10):2895–2903.

Londono, A. O. (2010). *Separation of Succinic Acid from Fermentation Broths and Esterification by a Reactive Distillation Method.* PhD thesis, Chemical Engineering, Michigan State University, United States of America.

López-Garzón, C. S. and Straathof, A. J. (2014). Recovery of carboxylic acids produced by fermentation. *Biotechnology Advances*, 32(5):873–904.

Lozowski, D. (2015). Economic indicators: Chemical engineering plant cost index. *Chemical Engineering and Processing: Process Intensification*, 122(5):104.

Luo, L., van der Voet, E., Huppes, G., and Udo de Haes,Helias A. (2009). Allocation issues in LCA methodology: a case study of corn stover-based fuel ethanol. *The International Journal of Life Cycle Assessment*, 14(6):529–539.

Luska, K. (2016). Pentane extraction scenarios - recent results from laboratory. *Personal communication*, Institute of Technical and Macromolecular Chemistry, Aachen, Germany.

Luska, K. L., Julis, J., Stavitski, E., Zakharov, D. N., Adams, A., and Leitner, W. (2014). Bifunctional nanoparticle–SILP catalysts (NPs@SILP) for the selective deoxygenation of biomass substrates. *Chemical Science*, 5(12):4895–4905.

Luterbacher, J. S., Rand, J. M., Alonso, D. M., Han, J., Youngquist, T. J., Maravelias, C. T., Pfleger, B. F., and Dumesic, J. A. (2014). Nonenzymatic Sugar Production from Biomass Using Biomass-Derived Gammavalerolactone. *Science*, 343(6168):277–280.

Ma, C., Wang, A., Qin, J., Li, L., Ai, X., Jiang, T., Tang, H., and Xu, P. (2009). Enhanced 2,3-butanediol production by *Klebsiella pneumoniae* SDM. *Applied Microbiology and Biotechnology*, 82(1):49–57.

Mahmudi, H. and Flynn, P. C. (2006). Rail vs truck transport of biomass. *Applied Biochemistry and Biotechnology*, 129-132:88–103.

Malueg, D. A. (1994). Monopoly output and welfare: The role of curvature of the demand function. *The Journal of Economic Education*, 25(3):235–250.

Manonmani, H. and Sreekantiah, K. (1987). Saccharification of sugar-cane bagasse with enzymes from *Aspergillus ustus* and *Trichoderma viride*. *Enzyme and Microbial Technology*, 9(8):484–488.

Mansoornejad, B., Chambost, V., and Stuart, P. (2010). Integrating product portfolio design and supply chain design for the forest biorefinery. *Computers & Chemical Engineering*, 34(9):1497–1506.

Mansoornejad, B., Pistikopoulos, E. N., and Stuart, P. (2013). Metrics for evaluating the forest biorefinery supply chain performance. *Computers & Chemical Engineering*, 54:125–139.

Mansoornejad, B., Stuart, P., and Pistikopolous, E. N. (2011). Incorporating flexibility design into supply chain design for forest biorefinery. *Journal of Science & Technology for Forest Products and Processes*, 1(2):54–66.

Mantau, U. (2012). Holzrohstoffbilanz Deutschland - Entwicklungen und Szenarien des Holzaufkommens und der Holzverwendung von 1987 bis 2015, http://www.saegeindustrie.de/tmp_uploads/00_holzrohstoffbilanz_2012.pdf, accessed July 6, 2016: Technical Report.

Manzer, L. E. (2010). Method of making 2-butanol, US Patent No. 2010/0056832 A1.

Marcotullio, G. and de Jong, W. (2010). Chloride ions enhance furfural formation from d-xylose in dilute aqueous acidic solutions. *Green Chemistry*, 12(10):1739–1746.

Marrero, J. and Gani, R. (2001). Group-contribution based estimation of pure component properties. *Fluid Phase Equilibria*, 183-184:183–208.

Martín, M. and Grossmann, I. E. (2013). On the systematic synthesis of sustainable biorefineries. *Industrial & Engineering Chemistry Research*, 52(9):3044–3064.

Marvin, W. A., Rangarajan, S., and Daoutidis, P. (2013a). Automated generation and optimal selection of biofuel-gasoline blends and their synthesis routes. *Energy & Fuels*, 27(6):3585–3594.

Marvin, W. A., Schmidt, L. D., and Daoutidis, P. (2013b). Biorefinery location and technology selection through supply chain optimization. *Industrial & Engineering Chemistry Research*, 52(9):3192–3208.

MathWorks (2016). Matlab R2016a, https://de.mathworks.com/products/matlab.html, accessed December, 2016.

Mavrovouniotis, M. L. (1991). Estimation of standard Gibbs energy changes of biotransformations. *The Journal of Biological Chemistry*, 266(22):14440–14445.

McElroy, C. R., Constantinou, A., Jones, L. C., Summerton, L., and Clark, J. H. (2015). Towards a holistic approach to metrics for the 21st century pharmaceutical industry. *Green Chemistry*, 17(5):3111–3121.

McKinlay, J. B., Shachar-Hill, Y., Zeikus, J. G., and Vieille, C. (2007). Determining *Actinobacillus succinogenes* metabolic pathways and fluxes by NMR and GC-MS analyses of 13C-labeled metabolic product isotopomers. *Metabolic Engineering*, 9(2):177–192.

Bibliography

Melin, T. and Rautenbach, R. (2007). *Membranverfahren: Grundlagen der Modul- und Anlagenauslegung*. VDI-Buch. Springer, Dordrecht.

Mergner, R., Janssen, R., Rutz, D., De Bari, I., and Sissot, F. (2013). Lignocellulosic ethanol process and demonstration: A handbook part I. *WIP Renewable Energies*, (München, Deutschland).

Meynial-Salles, I., Dorotyn, S., and Soucaille, P. (2008). A new process for the continuous production of succinic acid from glucose at high yield, titer, and productivity. *Biotechnology and Bioengineering*, 99(1):129–135.

Miettinen, K. (1998). *Nonlinear Multiobjective Optimization*, volume 12 of *International Series in Operations Research & Management Science*. Springer US, Boston, MA.

Millard, C. S., Chao, Y.-P., Liao, J. C., and Donnelly, M. I. (1996). Enhanced production of succinic acid by overexpression of *Phosphoenolpyruvate carboxylase* in *Escherichia coli*. *Applied and Environmental Microbiology*, 62(5):1808–1810.

Mitsos, A., Asprion, N., Floudas, C. A., Bortz, M., Baldea, M., Bonvin, D., Caspari, A., and Schäfer, P. (2018). Challenges in process optimization for new feedstocks and energy sources. *Computers & Chemical Engineering*, 113:209–221.

Mitsos, A., Bollas, G. M., and Barton, P. I. (2009). Bilevel optimization formulation for parameter estimation in liquid–liquid phase equilibrium problems. *Chemical Engineering Science*, 64(3):548–559.

Moncada, J., Posada, J. A., and Ramírez, A. (2015). Early sustainability assessment for potential configurations of integrated biorefineries. Screening of bio-based derivatives from platform chemicals. *Biofuels, Bioproducts and Biorefining*, 9(6):722–748.

Monigatti, L. (2016). Evaluation of the sustainability and economic viability of fuel production from renewable electricity. *Bachelor thesis (Supervisors: Ulonska, K., Bongartz, D.)*, Process Systems Engineering, RWTH Aachen University, Germany.

Morales-Rodriguez, R., Meyer, A. S., Gernaey, K. V., and Sin, G. (2011). Dynamic model-based evaluation of process configurations for integrated operation of hydrolysis and co-fermentation for bioethanol production from lignocellulose. *Bioresource Technology*, 102(2):1174–1184.

Nakamura, C. E. and Whited, G. M. (2003). Metabolic engineering for the microbial production of 1,3-propanediol. *Current Opinion in Biotechnology*, 14(5):454–459.

Neidleman, S. L., Amon, W. F., and Geigert, J. (1981). Process for the production of fructose, US Patent No. 4,246,347.

Ng, D. K. S., Tan, R. R., Foo, D. C., and El-Halwagi, M. M., editors (2016). *Process Design Strategies for Biomass Conversion Systems*. Wiley, United Kingdom, 1 edition.

Ng, L. Y., Andiappan, V., Chemmangattuvalappil, N. G., and Ng, D. K. (2015). A systematic methodology for optimal mixture design in an integrated biorefinery. *Computers & Chemical Engineering*, 81:288–309.

Ng, R. T., Kurniawan, D., Wang, H., Mariska, B., Wu, W., and Maravelias, C. T. (2018). Integrated framework for designing spatially explicit biofuel supply chains. *Applied Energy*, 216:116–131.

Ng, R. T. and Maravelias, C. T. (2017a). Design of biofuel supply chains with variable regional depot and biorefinery locations. *Renewable Energy*, 100:90–102.

Ng, R. T. and Maravelias, C. T. (2017b). Economic and energetic analysis of biofuel supply chains. *Applied Energy*, 205:1571–1582.

Nghiem, N. P., Davison, B. H., Suttle, B. E., and Richardson, G. R. (1997). Production of succinic acid by *Anaerobiospirillum succiniciproducens*. *Applied Biochemistry and Biotechnology*, 63-65(1):565–576.

Niziolek, A. M., Onel, O., Elia, J. A., Baliban, R. C., and Floudas, C. A. (2015). Coproduction of liquid transportation fuels and C 6 - C 8 aromatics from biomass and natural gas. *AIChE Journal*, 61(3):831–856.

Noureldin, N. A. and Lee, D. G. (1982). Heterogeneous permanganate oxidations. 2. Oxidation of alcohols using solid hydrated copper permanganate. *The Journal of Organic Chemistry*, 47(14):2790–2792.

Oh, I. J., Hye, W. L., Chul, H. P., Sang Yup, L., and Jinwon, L. (2008). Succinic acid production by continuous fermentation process using *Mannheimia succiniciproducens* LPK7. *Journal of Microbiology and Biotechnology*, 18(5):908–912.

Ohleyer, E., Blanch, H. W., and Wilke, C. R. (1985). Continuous production of lactic acid in a cell recycle reactor. *Applied Biochemistry and Biotechnology*, 11(4):317–332.

Okabe, M., Ohta, N., and Soo Park, Y. (1993). Itaconic acid production in an air-lift bioreactor using a modified draft tube. *Journal of Fermentation and Bioengineering*, 76(2):117–122.

Okino, S., Inui, M., and Yukawa, H. (2005). Production of organic acids by *Corynebacterium glutamicum* under oxygen deprivation. *Applied Microbiology and Biotechnology*, 68(4):475–480.

Okino, S., Noburyu, R., Suda, M., Jojima, T., Inui, M., and Yukawa, H. (2008). An efficient succinic acid production process in a metabolically engineered *Corynebacterium glutamicum* strain. *Applied Microbiology and Biotechnology*, 81(3):459–464.

Papadopoulos, A. I. and Linke, P. (2006). Multiobjective molecular design for integrated process-solvent systems synthesis. *AIChE Journal*, 52(3):1057–1070.

Papageorgiou, L. G. (2009). Supply chain optimisation for the process industries: Advances and opportunities. *Computers & Chemical Engineering*, 33(12):1931–1938.

Papaioannou, V., Adjiman, C. S., Jackson, G., and Galindo, A. (2011). Simultaneous prediction of vapour–liquid and liquid–liquid equilibria (VLE and LLE) of aqueous mixtures with the SAFT-gamma group contribution approach. *Fluid Phase Equilibria*, 306(1):82–96.

Papalexandri, K. P. and Pistikopoulos, E. N. (1996). Generalized modular representation framework for process synthesis. *AIChE Journal*, 42(4):1010–1032.

Park, Y. S., Itida, M., Ohta, N., and Okabe, M. (1994). Itaconic acid production using an air-lift bioreactor in repeated batch culture of *Aspergillus terreus*. *Journal of Fermentation and Bioengineering*, 77(3):329–331.

Park, Y. S., Ohta, N., and Okabe, M. (1993). Effect of dissolved oxygen concentration and impeller tip speed on itaconic acid production by *Aspergillus terreus*. *Biotechnology Letters*, 15(6):583–586.

Patel, A. D., Meesters, K., den Uil, H., de Jong, E., Blok, K., and Patel, M. K. (2012). Sustainability assessment of novel chemical processes at early stage: application to biobased processes. *Energy & Environmental Science*, 5(9):8430–8444.

Patel, M., Crank, M., Dornburg, V., Hermann, B., Roes, L., Hüsing, B., Overbeek, L., Terragni, F., and Recchia, E. (2006). Medium and Long-term Opportunities and Risks of the Biotechnological Production of Bulk Chemicals from Renewable Resources - The Potential of White Biotechnology: The BREW Project.

Penner, D., Redepenning, C., Mitsos, A., and Viell, J. (2017). Conceptual design of methyl ethyl ketone production via 2,3-butanediol for fuels and chemicals. *Industrial & Engineering Chemistry Research*, 56(14):3947–3957.

Pertsinidis, A. (1992). *On the Parametric Optimization of Mathematical Programs with Binary Variables and its Applications in Chemical Engineering Process Synthesis*. PhD thesis, Carnegie Mellon University, Pittsburgh.

Pertsinidis, A., Grossmann, I., and McRae, G. (1998). Parametric optimization of MILP programs and a framework for the parametric optimization of MINLPs. *Computers & Chemical Engineering*, 22:S205–S212.

Petley, G. J. (1997). A Method for Estimating the Capital Cost of Chemical Process Plants : fuzzy matching, https://dspace.lboro.ac.uk/2134/11165. PhD thesis, Chemical Engineering, Loughborough University, United Kingdom.

Pham, V. and El-Halwagi, M. (2012). Process synthesis and optimization of biorefinery configurations. *AIChE Journal*, 58(4):1212–1221.

Pistikopoulos, E. N. (1995). Uncertainty in process design and operations. *Computers & Chemical Engineering*, 19(1):553–563.

Poth, S. (2013). *Enzymatische Hydrolyse und Fermentation von Lignocellulose: Optimerung und Prozessintegration zur Umsetzung von vorbehandelten, hölzernen Cellulose-Faserstoffen für die Produktion von Ethanol*. PhD thesis, Technische Universität Kaiserslautern, Germany.

Potvin, J., Sorlien, E., Hegner, J., DeBoef, B., and Lucht, B. L. (2011). Effect of NaCl on the conversion of cellulose to glucose and levulinic acid via solid supported acid catalysis. *Tetrahedron Letters*, 52(44):5891–5893.

Pretel, E. J., López, P. A., Bottini, S. B., and Brignole, E. A. (1994). Computer-aided molecular design of solvents for separation processes. *AIChE Journal*, 40(8):1349–1360.

Pyrgakis, K. A. and Kokossis, A. C. (2016). A New Methodology to Apply Total Site Analysis as a Synthesis Tool to Select and Integrate Processes in Multiple-Product Biorefinery Plants. In Zdravko, K. and Milos, B., editors, *26th European Symposium on Computer Aided Process Engineering - ESCAPE 26*, volume 38, pages 2073–2078.

Quaglia, A., Gargalo, C. L., Chairakwongsa, S., Sin, G., and Gani, R. (2015). Systematic network synthesis and design: Problem formulation, superstructure generation, data management and solution. *Computers & Chemical Engineering*, 72:68–86.

Quaglia, A., Sarup, B., Sin, G., and Gani, R. (2013). A systematic framework for enterprise-wide optimization: Synthesis and design of processing networks under uncertainty. *Computers & Chemical Engineering*, 59:47–62.

Qureshi, N. and Cheryan, M. (1989). Production of 2,3-butanediol by *Klebsiella oxytoca*. *Appl Microbiol Biotechnology*, 30:440–443.

Rajagopalan, N., Venditti, R., Kelley, S., and Daystar, J. (2017). Multi-attribute uncertainty analysis of the life cycle of lignocellulosic feedstock for biofuel production. *Biofuels, Bioproducts and Biorefining*, 11(2):269–280.

Ramesh, M. (2013). Thermodynamics of metabolic networks. *Student thesis (Supervisor: Jonathan Meade, Michael M. Domach, Ulonska, K.)*, Carnegie Mellon University, Department of Chemical Engineering, USA and Process Systems Engineering, RWTH Aachen University,Germany.

Rangaiah, G. P. (2009). *Multi-objective Optimization: Techniques and Applications in Chemical Engineering*, volume 1 of *Advances in process systems engineering*. Hackensack, N.J. and World Scientific, Singapore.

Rangarajan, S., Bhan, A., and Daoutidis, P. (2012). Language-oriented rule-based reaction network generation and analysis: Description of RING. *Computers & Chemical Engineering*, 45:114–123.

Rangarajan, S., Bhan, A., and Daoutidis, P. (2014a). Identification and analysis of synthesis routes in complex catalytic reaction networks for biomass upgrading. *Applied Catalysis B: Environmental*, 145:149–160.

Rangarajan, S., Kaminski, T., van Wyk, E., Bhan, A., and Daoutidis, P. (2014b). Language-oriented rule-based reaction network generation and analysis: Algorithms of RING. *Computers & Chemical Engineering*, 64:124–137.

Recker, S. (2017). *Systematic and Optimization-Based Synthesis and Design of Chemical Processes*. PhD thesis, Process Systems Engineering, Rheinisch-Westfälische Technische Hochschule Aachen, Germany.

Redepenning, C. and Marquardt, W. (2017). Pinch-based shortcut method for the conceptual design of adiabatic absorption columns. *AIChE Journal*, 63(4):1213–1225.

Rekkas Ventiris, M. (2015). Conceptual design for an early stage process analysis of reaction and separation processes in future biorefineries. *Diploma thesis (Supervisor: Ulonska, K.)*, Process Systems Engineering, RWTH Aachen University,Germany and National Technical University of Athens.

Ren, J., Xu, D., Cao, H., Wei, S., Goodsite, M. E., and Dong, L. (2015). Sustainability Decision Support Framework for Industrial System Prioritization. *AIChE Journal*, page n/a.

Renewable Fuels Association (2017). World fuel ethanol production, http://www.ethanolrfa.org/resources/industry/statistics/#1454099103927-61e598f7-7643, accessed September, 19, 2017.

Ribeiro, M. G. T. and Machado, A. A. (2013). Greenness of chemical reactions – limitations of mass metrics. *Green Chemistry Letters and Reviews*, 6(1):1–18.

Rizwan, M., Lee, J. H., and Gani, R. (2015). Optimal design of microalgae-based biorefinery: Economics, opportunities and challenges. *Applied Energy*, 150:69–79.

Ruth, M. (2011). Hydrogen Production Cost Estimate Using Biomass Gasification: Independent Review: Technical Report DE-AC36-08GO28308 National Renewable Energy Laboratory.

S and P Global (2016). Methyl ethyl ketone spot price- Release date October 2014, http://www.platts.com/latest-news/petrochemicals/london/methyl-ethyl-ketone-spot-prices-skyrocket-in-26911566, accessed November 23, 2016.

Sahinidis, N. V. (2004). Optimization under uncertainty: state-of-the-art and opportunities. *Computers & Chemical Engineering*, 28(6-7):971–983.

Saint-Amans, S., Perlot, P., Goma, G., and Soucaille, P. (1994). High production of 1,3-propanediol from glycerol by *Clostridium butyricum* VPI 3266 in a simply controlled fed-batch system. *Biotechnology Letters*, 16(8):831–836.

Saling, P., Kicherer, A., Dittrich-Krämer, B., Wittlinger, W., Zombik, W., Schmidt, I., Schrott, W., and Schmidt, S. (2002). Eco-efficiency analysis by BASF: the method. *The International Journal of Life Cycle Assessment*, 7(4):203–218.

Saltelli, A., Ratto, M., Andres, T., Campolongo, F., Cariboni, J., Gatelli, D., Saisana, M., and Tarantola, S. (2007). *Global Sensitivity Analysis: The Primer*. John Wiley & Sons Inc., Hoboken, USA.

Saltelli, A., Tarantola, S., Campolongo, F., and Ratto, M. (2004). *Sensitivity Analysis in Practice: A Guide to Assessing Scientific Models*, volume 20859. Wiley, Hoboken, USA.

Sammons, N., Yuan, W., Eden, M., Aksoy, B., and Cullinan, H. (2008). Optimal biorefinery product allocation by combining process and economic modeling. *Chemical Engineering Research and Design*, 86(7):800–808.

Samuelov, N. S., Lamed, R., Lowe, S., and Zeikus, J. G. (1991). Influence of CO(2)-HCO(3)(−) levels and pH on growth, succinate production, and enzyme activities of *Anaerobiospirillum succiniciproducens*. *Applied and Environmental Microbiology*, 57(10):3013–3019.

Sánchez, A. M., Bennett, G. N., and San, K.-Y. (2005). Novel pathway engineering design of the anaerobic central metabolic pathway in *Escherichia coli* to increase succinate yield and productivity. *Metabolic Engineering*, 7(3):229–239.

Sandin, G., Røyne, F., Berlin, J., Peters, G. M., and Svanström, M. (2015). Allocation in LCAs of biorefinery products: implications for results and decision-making. *Journal of Cleaner Production*, 93:213–221.

Santibanez-Aguilar, J. E., Gonzalez-Campos, J. B., Ponce-Ortega, J. M., Serna-Gonzalez, M., and El-Halwagi, M. M. (2011). Optimal planning of a biomass conversion system considering economic and environmental aspects. *Industrial & Engineering Chemistry Research*, 50(14):8558–8570.

Sauer, M., Porro, D., Mattanovich, D., and Branduardi, P. (2008). Microbial production of organic acids: expanding the markets. *Trends in biotechnology*, 26(2):100–108.

Scheffczyk, J., Redepenning, C., Jens, C. M., Winter, B., Leonhard, K., Marquardt, W., and Bardow, A. (2016). Massive, automated solvent screening for minimum energy demand in hybrid extraction–distillation using COSMO-RS. *Chemical Engineering Research and Design*, 115:433–442.

Schilling, C. H. and Palsson, B. O. (1998). The underlying pathway structure of biochemical reaction networks. *Proceedings of the National Academy of Sciences of United States of America*, 95:4193–4198.

Schwaderer, F. (2012). *Integrierte Standort-, Kapazitäts- und Technologieplanung von Wertschöpfungsnetzwerken zur stofflichen und energetischen Biomassenutzung*. PhD thesis, Wirtschaftswissenschaften, Karlsruher Institut für Technologie, Germany.

Searcy, E., Flynn, P., Ghafoori, E., and Kumar, A. (2007). The relative cost of biomass energy transport. *Applied Biochemistry and Biotechnology*, 137-140(1-12):639–652.

Sepiacci, P., Depetri, V., and Manca, D. (2017). A systematic approach to the optimal design of chemical plants with waste reduction and market uncertainty. *Computers & Chemical Engineering*, 102:96–109.

Serrano-Ruiz, J. C. and Dumesic, J. A. (2011). Catalytic routes for the conversion of biomass into liquid hydrocarbon transportation fuels. *Energy & Environmental Science*, 4(1):83–99.

Serrano-Ruiz, J. C., West, R. M., and Dumesic, J. A. (2010). Catalytic conversion of renewable biomass resources to fuels and chemicals. *Annual Review of Chemical and Biomolecular Engineering*, 1(1):79–100.

Shabani, N., Akhtari, S., and Sowlati, T. (2013). Value chain optimization of forest biomass for bioenergy production: A review. *Renewable and Sustainable Energy Reviews*, 23:299–311.

Shah, N. (2005). Process industry supply chains: Advances and challenges. *Computers & Chemical Engineering*, 29(6):1225–1235.

Shanghai Polymet Commodities Ltd. (2016). FOB price for dichloromethane, http://www.alibaba.com/product-detail/dichloromethane_370296049.html?spm=a2700. 7724857.29.21.93y0E7, accessed January 8, 2016.

Sharifzadeh, M., Garcia, M. C., and Shah, N. (2015). Supply chain network design and operation: Systematic decision-making for centralized, distributed, and mobile biofuel production using mixed integer linear programming (MILP) under uncertainty. *Biomass and Bioenergy*, 81:401–414.

Sharma, P., Romagnoli, J., and Vlosky, R. (2013). Options analysis for long-term capacity design and operation of a lignocellulosic biomass refinery. *Computers & Chemical Engineering*, 58:178–202.

Sholl, D. S. and Lively, R. P. (2016). Seven chemical separations to change the world. *Nature*, 532(7600):435–437.

Simangunsong, B. C. H. and Buongiorno, J. (2001). International demand equations for forest products: A comparison of methods. *Scandinavian Journal of Forest Research*, 16(2):155–172.

Sinnott, R. K., Coulson, J. M., and Richardson, J. F. (2005). *Chemical Engineering Design*, volume 6 of *Coulson & Richardson's chemical engineering*. Elsevier Butterworth-Heinemann, Oxford, 4th edition.

Siougkrou, E. and Kokossis, A. (2016a). Development of semantically-enabled community hubs in biorefineries and biorenewables. In Zdravko, K. and Milos, B., editors, *26th European Symposium on Computer Aided Process Engineering - ESCAPE 26*, volume 38, pages 2013–2018.

Siougkrou, E. and Kokossis, A. (2016b). IPSEN Tools, http://tools.ipsen.ntua.gr/ipsentools/web/index.php, accessed December 5, 2016.

SIX Financial Information (2016). Exchange rate USD in Euro, http://www.finanzen.net/waehrungsrechner/us-dollar_euro, accessed October 25, 2016.

Skiborowski, M. (2014). *Optimization-Based Methods for the Conceptual Design of Separation Processes for Azeotropic Mixtures*. PhD thesis, Process Systems Engineering, Rheinisch-Westfälische Technische Hochschule Aachen.

Bibliography

Skiborowski, M., Harwardt, A., and Marquardt, W. (2013). Conceptual design of distillation-based hybrid separation processes. *Annual Review of Chemical and Biomolecular Engineering*, 4(1):45–68.

Skiborowski, M., Harwardt, A., and Marquardt, W. (2015). Efficient optimization-based design for the separation of heterogeneous azeotropic mixtures. *Computers & Chemical Engineering*, 72:34–51.

Smith, K. M. and Liao, J. C. (2011). An evolutionary strategy for isobutanol production strain development in *Escherichia coli*. *Metabolic Engineering*, 13(6):674–681.

Smith, R. and Linnhoff, B. (1988). The design of separators in the context of overall processes. *Chemical Engineering Research and Design*, 66(3):195–228.

Song, H., Jang, S. H., Park, J. M., and Lee, S. Y. (2008). Modeling of batch fermentation kinetics for succinic acid production by *Mannheimia succiniciproducens*. *Biochemical Engineering Journal*, 40(1):107–115.

Song, H., Lee, J. W., Choi, S., You, J. K., Hong, W. H., and Lee, S. Y. (2007). Effects of dissolved CO_2 levels on the growth of *Mannheimia succiniciproducens* and succinic acid production. *Biotechnology and bioengineering*, 98(6):1296–1304.

Sorda, G. and Madlener, R. (2012). Cost-effectiveness of lignocellulose biorefineries and their impact on the deciduous wood markets in Germany, http://www.fcn.eonerc.rwth-aachen.de/go/id/dnfj/file/206264.

Srinivas, B. and El-Halwagi, M. (1994). Synthesis of combined heat and reactive mass-exchange networks. *Chemical Engineering Science*, 49(13):2059–2074.

Steimel, J. and Engell, S. (2016). Optimization-based support for process design under uncertainty: A case study. *AIChE Journal*, 62(9):3404–3419.

Stephen, H. and Stephen, T. (2013). *Solubilities of inorganic and organic compounds*. Ebrary online. Pergamon Press, Oxford and New York and Toronto and Sydney and Paris and Frankfurt, reprinted. edition.

Storch, M., Hinrichsen, F., Wensing, M., Will, S., and Zigan, L. (2015). The effect of ethanol blending on mixture formation, combustion and soot emission studied in an optical DISI engine. *Applied Energy*, 156:783–792.

Straathof, A. J. J. (2014). Transformation of Biomass into Commodity Chemicals Using Enzymes or Cells. *Chemical Reviews*, 114(3):1871–1908.

Sy, C. L., Ubando, A. T., Aviso, K. B., and Tan, R. R. (2018). Multi-objective target oriented robust optimization for the design of an integrated biorefinery. *Journal of Cleaner Production*, 170:496–509.

Szmant, H. H. and Chundury, D. D. (1981). The preparation of 5-hydroxymethylfurfuraldehyde from high fructose corn syrup and other carbohydrates. *Journal of Chemical Technology and Biotechnology*, 31(1):131–145.

Tamura, M., Tokonami, K., Nakagawa, Y., and Tomishige, K. (2013). Rapid synthesis of unsaturated alcohols under mild conditions by highly selective hydrogenation. *Chemical Communications*, 49(63):7034.

Tang, M. C., Chin, M. W. S., Lim, K. M., Mun, Y. S., Ng, R. T. L., Tay, D. H. S., and Ng, D. K. S. (2013). Systematic approach for conceptual design of an integrated biorefinery with uncertainties. *Clean Technologies and Environmental Policy*, 15(5):783–799.

Tao, L., Tan, E. C. D., McCormick, R., Zhang, M., Aden, A., He, X., and Zigler, B. T. (2014). Techno-economic analysis and life-cycle assessment of cellulosic isobutanol and comparison with cellulosic ethanol and n-butanol. *Biofuels, Bioproducts and Biorefining*, 8(1):30–48.

Tawarmalani, M. and Sahinidis, N. V. (2005). A polyhedral branch-and-cut approach to global optimization. *Mathematical Programming*, 103(2):225–249.

The DOW Chemical Company (2018). Answer to "Registration of isobutanol as a gasoline additive: opportunity for public comment", https://www.regulations.gov/document?D=EPA-HQ-OAR-2018-0131-0001, accessed June, 05, 2018.

Thewes, M., Muther, M., Brassat, A., Pischinger, S., and Sehr, A. (2012). Analysis of the effect of bio-fuels on the combustion in a downsized DI SI engine. *SAE International Journal of Fuels and Lubricants*, 5(1):274–288.

Thilagavathi, N. and Jayabalakrishnan, C. (2010). Synthesis, characterization, catalytic and antimicrobial studies of ruthenium(III) complexes. *Open Chemistry*, 8(4):842–851.

Thomas, K. and Ingledew, W. (1992). Production of 21% (v/v) ethanol by fermentation of very high gravity (VHG) wheat mashes. *Journal of Industrial Microbiology*, 10:61–68.

Tock, L. and Maréchal, F. (2012). Co-production of hydrogen and electricity from lignocellulosic biomass: Process design and thermo-economic optimization. *Energy*, 45(1):339–349.

Tock, L. and Maréchal, F. (2015). Decision support for ranking Pareto optimal process designs under uncertain market conditions. *Computers & Chemical Engineering*, 83:165–175.

Bibliography

Transparency Market Research (2016). Butanediol, Butadiene and MEK market - Release date October 2012, http://www.transparencymarketresearch.com/butanediol-butadiene-and-mek-market.html, accessed November 23, 2016.

Trost, B. M. (1991). The atom economy: A search for synthetic efficiency. *Science*, 254(5037):1471–1477.

Tsagkari, M., Couturier, J.-L., Kokossis, A., and Dubois, J.-L. (2016). Early-stage capital cost estimation of biorefinery processes: A comparative study of heuristic techniques. *ChemSusChem*, 9(17):2284–2297.

Tsakalova, M., Yang, A., and Kokossis, A. C. (2014). A Systems Approach for the Holistic Screening of Second Generation Biorefinery Paths for Energy and Biobased Products. In Klemeš, J. J., Varbanov, P. S., and Liew, P. Y., editors, *24th European Symposium on Computer Aided Process Engineering*, Computer aided chemical engineering, pages 205–210. Elsevier, Amsterdam and Oxford.

Tula, A. K., Babi, D. K., Bottlaender, J., Eden, M. R., and Gani, R. (2017). A computer-aided software-tool for sustainable process synthesis-intensification. *Computers & Chemical Engineering*, 105:74–95.

Uhlman, B. W. and Saling, P. (2010). Measuring and communicating sustainability through eco-efficiency analysis. *Chemical Engineering Progress*, pages 17–26.

UK Forex Foreign Exchange (2016). Exchange rate Great Britain Pound into United States Dollar in 2000, accessed September 22, 2016.

Ulonska, K., Ebert, B. E., Blank, L. M., Mitsos, A., and Viell, J. (2015). Systematic screening of fermentation products as future platform chemicals for biofuels. In Gernaey, K. V., editor, *12th International Symposium on Process Systems Engineering and 25th European Symposium on Computer Aided Process Engineering*, pages 1331–1336.

Ulonska, K., König, A., Klatt, M., Mitsos, A., and Viell, J. (2018). Optimization of multiproduct biorefinery processes under consideration of biomass supply chain management and market developments. *Industrial & Engineering Chemistry Research*, 57(20):6980–6991.

Ulonska, K., Skiborowski, M., Mitsos, A., and Viell, J. (2016a). Early-stage evaluation of biorefinery processing pathways using process network flux analysis. *AIChE Journal*, 62(9):3096–3108.

Ulonska, K., Voll, A., and Marquardt, W. (2016b). Screening pathways for the production of next generation biofuels. *Energy & Fuels*, 30(1):445–456.

Urbance, S. E., Pometto, A. L., Dispirito, A. A., and Denli, Y. (2004). Evaluation of succinic acid continuous and repeat-batch biofilm fermentation by *Actinobacillus succinogenes* using plastic composite support bioreactors. *Applied Microbiology and Biotechnology*, 65(6):664–670.

Urbanus, J., Roelands, C., Verdoes, D., and ter Horst, J. H. (2012). Intensified crystallization in complex media: Heuristics for crystallization of platform chemicals. *Chemical Engineering Science*, 77:18–25.

Valdivia, M., Galan, J. L., Laffarga, J., and Ramos, J.-L. (2016). Biofuels 2020: Biorefineries based on lignocellulosic materials. *Microbial biotechnology*, 9(5):585–594.

Vane, L. M. (2008). Separation technologies for the recovery and dehydration of alcohols from fermentation broths. *Biofuels, Bioproducts and Biorefining*, 2(6):553–588.

Vellguth, A. (2016). Assessing the sustainability and efficiency of bio-based fuels. *Bachelor thesis (Supervisor: Ulonska, K.)*, Process Systems Engineering, RWTH Aachen University, Germany.

Vemuri, G. N., Eiteman, M. A., and Altman, E. (2002). Succinate production in dual-phase *Escherichia coli* fermentations depends on the time of transition from aerobic to anaerobic conditions. *Journal of industrial microbiology & biotechnology*, 28(6):325–332.

Verkehrsrundschau (2016). Verkehrsrundschau-Index - Preisindex für den Straßengüterverkehr in Deutschland, http://www.verkehrsrundschau.de/vr-index-lkw-frachtraten-unter-druck-1837126.html, accessed June 1, 2016.

Victoria Villeda, J. (2016). *Reaction Network Generation and Evaluation for the Design of Biofuel Value Chains*. PhD thesis, Process Systems Engineering, Rheinisch-Westfälische Technische Hochschule Aachen, Germany.

Viell, J., Harwardt, A., Seiler, J., and Marquardt, W. (2013). Is biomass fractionation by Organosolv-like processes economically viable? A conceptual design study. *Bioresource Technology*, 150:89–97.

Voll, A. (2014). *Model-Based Screening of Reaction Pathways for Processing of Biorenewables*. PhD thesis, Process Systems Engineering, Rheinisch-Westfälische Technische Hochschule Aachen, Germany.

Voll, A. and Marquardt, W. (2012a). Benchmarking of next-generation biofuels from a process perspective. *Biofuels Bioproducts & Biorefining*, 6(3):292–301.

Voll, A. and Marquardt, W. (2012b). Reaction network flux analysis: Optimization-based evaluation of reaction pathways for biorenewables processing. *AIChE Journal*, 58(6):1788–1801.

Wang, B., Gebreslassie, B. H., and You, F. (2013). Sustainable design and synthesis of hydrocarbon biorefinery via gasification pathway: Integrated life cycle assessment and technoeconomic analysis with multiobjective superstructure optimization. *Computers & Chemical Engineering*, 52:55–76.

Weiss, J. M. and Downs, C. R. (1923). The physical properties of maleic, fumaric and malic acids. *Journal of the American Chemical Society*, 45(4):1003–1008.

Welter, K. (2000). *Biotechnische Produktion von Itaconsäure aus nachwachsenden Rohstoffen mit immobilisierten Zellen*. PhD thesis, Naturwissenschaftliche Fakultät, Technische Universität Carolo-Wilhelmina zu Braunschweig, Germany.

Werpy, T. and Petersen, G. (2004). Top Value Added Chemicals from Biomass: Volume I-Results of Screening for Potential Candidates from Sugars and Synthesis Gas: Technical Report DOE/GO-102004-1992 U.S. Department of Energy.

West, R. M., Liu, Z. Y., Peter, M., Gärtner, C. A., and Dumesic, J. A. (2008). Carbon–carbon bond formation for biomass-derived furfurals and ketones by aldol condensation in a biphasic system. *Journal of Molecular Catalysis A: Chemical*, 296(1-2):18–27.

Westerberg, A. W. (2004). A retrospective on design and process synthesis. *Computers & Chemical Engineering*, 28(4):447–458.

Wewetzer, S. (2014). Itaconic acid prodction using *Ustilago maydis*. *Personal communication*, Institute of Biochemical Engineering, Aachen, Germany.

Wittig, R., Lohmann, J., and Gmehling, J. (2003). Vapor−liquid equilibria by UNIFAC group contribution. 6. revision and extension. *Industrial & Engineering Chemistry Research*, 42(1):183–188.

Wloemer, M. (2016). Conceptual design of a biorefinery for ethanol production. *Master thesis (Supervisors: Ulonska, K., Recker, S.)*, Process Systems Engineering, RWTH Aachen University, Germany.

Woolston, B. M., Edgar, S., and Stephanopoulos, G. (2013). Metabolic engineering: past and future. *Annual Review of Chemical and Biomolecular Engineering*, 4(1):259–288.

Wu, W., Henao, C. A., and Maravelias, C. T. (2016). A superstructure representation, generation, and modeling framework for chemical process synthesis. *AIChE Journal*, 62(9):3199–3214.

Wuhan Benjamin Pharmaceutical Chemical Co. Ltd (2016). FOB price for 2-Methyltetrahydrofuran, http://www.alibaba.com/product-detail/96-47-9-2-Methyltetrahydrofuran_1920680121.html?spm=a2700.7724838.30.157.JBNcME, accessed January 8, 2016.

Yahiro, K., Takahama, T., Park, Y. S., and Okabe, M. (1995). Breeding of *Aspergillus terreus* mutant TN-484 for itaconic acid production with high yield. *Journal of Fermentation and Bioengineering*, 79(5):506–508.

Yamane, T. and Tanaka, R. (2013). Highly accumulative production of L(+)-lactate from glucose by crystallization fermentation with immobilized *Rhizopus oryzae*. *Journal of bioscience and bioengineering*, 115(1):90–95.

Yang, Z., Huang, Y.-B., Guo, Q.-X., and Fu, Y. (2013). Raney Ni catalyzed transfer hydrogenation of levulinate esters to gamma-valerolactone at room temperature. *Chemical Communications*, 49(46):5328.

Yanowitz, J., Christensen, E., and McCormick, R. L. (2011). Utilization of Renewable Oxygenates as Gasoline Blend Components: Technical Report NREL/TP-5400-50791 National Renewable Energy Laboratory.

Yee, T. and Grossmann, I. (1990). Simultaneous optimization models for heat integration—II. Heat exchanger network synthesis. *Computers & Chemical Engineering*, 14(10):1165–1184.

Yenkie, K. M., Wu, W., Clark, R. L., Pfleger, B. F., Root, T. W., and Maravelias, C. T. (2016). A roadmap for the synthesis of separation networks for the recovery of bio-based chemicals: Matching biological and process feasibility. *Biotechnology Advances*, 34(8):1362–1383.

Yeomans, H. and Grossmann, I. E. (1999). A systematic modeling framework of superstructure optimization in process synthesis. *Computers & Chemical Engineering*, 23(6):709–731.

Yim, H., Haselbeck, R., Niu, W., Pujol-Baxley, C., Burgard, A., Boldt, J., Khandurina, J., Trawick, J. D., Osterhout, R. E., Stephen, R., Estadilla, J., Teisan, S., Schreyer, H. B., Andrae, S., Yang, T. H., Lee, S. Y., Burk, M. J., and van Dien, S. (2011). Metabolic engineering of *Escherichia coli* for direct production of 1,4-butanediol. *Nature Chemical Biology*, 7(7):445–452.

You, F., Tao, L., Graziano, D. J., and Snyder, S. W. (2012). Optimal design of sustainable cellulosic biofuel supply chains: Multiobjective optimization coupled with life cycle assessment and input-output analysis. *AIChE Journal*, 58(4):1157–1180.

Bibliography

You, F. and Wang, B. (2011). Life cycle optimization of biomass-to-liquid supply chains with distributed–centralized processing networks. *Industrial & Engineering Chemistry Research*, 50(17):10102–10127.

Yuan, Y., Leng, Y., Shao, H., Huang, C., and Shan, K. (2014). Solubility of DL-malic acid in water, ethanol and in mixtures of ethanol+water. *Fluid Phase Equilibria*, 377:27–32.

Zauba Technologies and Data Services Pvt Ltd. (2015). Detailed Import Data of gamma butyrolactone, https://www.zauba.com/import-gamma-butyrolactone-hs-code.html, accessed November 4, 2015.

Zeng, A.-P., Biebl, H., and Deckwer, W.-D. (1991). Production of 2,3-butanediol in a membrane bioreactor with cell recycle. *Applied Microbiology and Biotechnology*, 34(4):463–468.

Zevnik, F. C. and Buchanan, R. L. (1963). Generalized correlation for process investment. *Chemical Engineering Progress*, 59(2):70–77.

Zhang, D., del Rio-Chanona, E. A., and Shah, N. (2017). Screening synthesis pathways for biomass-derived sustainable polymer production. *ACS Sustainable Chemistry & Engineering*, 5(5):4388–4398.

Zhang, W., Yu, D., Ji, X., and Huang, H. (2012). Efficient dehydration of bio-based 2,3-butanediol to butanone over boric acid modified HZSM-5 zeolites. *Green Chemistry*, 14(12):3441–3450.

Zondervan, E., Nawaz, M., de Haan, A. B., Woodley, J. M., and Gani, R. (2011). Optimal design of a multi-product biorefinery system. *Computers & Chemical Engineering*, 35(9):1752–1766.